The SCIENCE of
EVERYDAY LIFE

MARTY JOPSON

The SCIENCE of

EVERYDAY LIFE

Why teapots dribble, toast burns
and light bulbs shine

Michael O'Mara Books Limited

First published in Great Britain in 2015 by
Michael O'Mara Books Limited
9 Lion Yard
Tremadoc Road
London SW4 7NQ

A CIP catalogue record for this book is available from the British Library.

Papers used by Michael O'Mara Books Limited are natural, recyclable
products made from wood grown in sustainable forests. The
manufacturing processes conform to the environmental regulations of the
country of origin.

ISBN: 978-1-78243-418-4 in hardback print format
ISBN: 978-1-78243-420-7 in ebook format

2 3 4 5 6 7 8 9 10

Jacket design by www.shepherdstudio.co.uk
Illustrations by Emma McGowan
Designed and typeset by K DESIGN, Winscombe, Somerset

Printed and bound by CPI Group (UK) Ltd, Croydon, CR0 4YY

www.mombooks.com

For Juliet, Poppy and George,
all of whom suffer from insatiable curiosity.

Contents

Introduction

All around us, every day, we are surrounded by some of the most interesting examples of science. Yet we don't see it. The science is hidden in plain sight, submerged beneath the everyday phenomena of our lives and the hand-held gizmos we take for granted. The best bits of science are going unnoticed.

If you pause for a moment and scratch the surface, though, the gleam of fascinating science shines through. Take, for example, hot, spicy foods and what actually causes the sensation of heat. The effect of the causative molecule in chillies, capsaicin, has been studied right down to the molecular level, but that is only the start of the spicy story. There are also spicy molecules such as piperine, gingerol, allyl isothiocyanate and the tongue-twisting and tongue-numbing hydroxy-alpha-sanshool. Each molecule is different, yet each directly affects our nerves to mimic pain.

Even the most mundane of technological devices that we give barely a moment's attention can harbour fascinating science. When was the last time you considered what was really going on inside a quartz clock? Yet the feedback electrical system used to set up the oscillations within a quartz crystal is fiendishly ingenious, and it's not just in clocks, it's inside every mobile phone, computer and tablet device. Or there are the infrared motion sensors, part of every alarm system that silently watches us all as we go about our lives. What makes them so brilliant is hidden within; the cunning wiring of two tiny crystals allowing the sensor to not only see parts of the spectrum that we cannot, but to react only to moving sources of infrared light above a certain size.

In many cases, once you get past the initial understanding, you find yourself abruptly at the scientific coalface, where the ultimate answer is that we just don't know the answer – so far. From the possibility of mining our road sweepings for precious platinum, to what makes a moth circle a light, the science of everyday life still contains unexplored territory.

However, does understanding this everyday science matter? On the surface the answer to this is no. Superficially, it doesn't make any difference in the slightest if you don't know how a toaster works, or why sitting under a tree is delightfully cooling. These things still happen even in the face of ignorance, but knowing why can make a difference.

Crucially, in our technologically controlled world, greater understanding can lead to better-informed decisions. These might be small but vital choices, such as what sort of object to poke into a toaster to free a stuck piece of bread. Knowing that the bare nichrome wires in a toaster have mains electricity flowing in them should convince you to use a wooden spoon or a chopstick, rather than an electrically conducting metal knife. By understanding the workings of a simple item like a toaster you can shape your interactions with it, increasing its usefulness and function. Similarly, appreciating the cooling effect of green vegetation provides an informed reason to choose to plant green spaces within our cities and towns.

It's not just about the appliance of science for the creation of handy life hacks and city planning; there is a more nebulous, but fundamental reason why everyday science is important. It makes life more exciting. Knowing the context and explanation for something makes the experience of the thing vastly more enriching. No one would deny that this is true for a great work of art or literature, and the same is true for science. Once you know why your fingers really go wrinkly in the bathtub, you will never look at your prune-like fingertips in the same way again. Your bath time just became more interesting.

This book sets out to reveal some of the astonishing and intriguing science going on all around you all the time. What's more, it's not as if this is old science that has been long established. To put yourself at the cutting edge of scientific understanding you don't need to travel to the extremes of our world, into deep space, or to collide subatomic particles at nearly the speed of light. All you need do is look around you and dive into the subtlety and complexity of the science of your own everyday life.

The Sustaining Science of Food and Drink

The sweetest thing

Sweet, juicy strawberries, cake still warm from the oven and, my favourite, honey straight from the comb... Most of us enjoy eating sweet things, to the extent that seeking them out seems to be hard-wired into our brains. Yet our ability to taste sweetness is remarkably non-specific, and is fooled by a host of chemicals that seem to bear little resemblance to sugar. Not only that, but when it comes to sweetness, ordinary sugar, or sucrose, isn't very sweet at all.

The sweetest chemical so far discovered goes by the name of lugduname and ranks about 250,000 times sweeter than sucrose. What's perplexing for chemists, though, is that lugduname doesn't bear any structural resemblance to other sugars. This poses a little bit of a problem for science, as the usual way that a chemical receptor works is that it recognizes just a small portion of a molecule, maybe the arrangement of a half-dozen atoms or so. It doesn't matter what shape the rest of the molecule is, as long as those half-dozen atoms are in the right places. It's called the lock and key model, and so long as a chemical has the key, it will fit the lock. Sucrose and lugduname don't appear to share any such kind of key.

The term sugar itself denotes a group of chemicals of different lengths of chains of carbon atoms, including an oxygen, and often bent into a ring. The simplest sugars contain just one of these rings, and include glucose and fructose. Two simple sugars can hook together to make compounds such as sucrose, which is really a fructose stuck together with a glucose. All of these chemicals share common structures, and it is easy to imagine how it is that they register as sweet, as they all possess the right key.

Things start to get a bit weirder when you look at sugar substitutes. We are all familiar with sweeteners, such as aspartame, found in a whole host of food products including diet fizzy drinks. Many people presume that sugar substitutes are entirely synthetic and made in a lab. It turns out that nature was there long before the diet industry, and you can find sugar substitutes in surprising places.

My own personal favourite, because it surprised me when I first encountered it on an ecology field trip, can be found at the seaside. Next time you walk along a rocky shore line keep your eyes peeled for fronds of *Saccharina latissima*, or sugar kelp as it is commonly known. It's fairly distinctive and easy to spot once you know what to look for. It's a type of brown seaweed that comes in single, undivided blades, and is often a couple of metres (around 6 ft) long and about 10 to 15 cm (4–6 inches) wide. What makes it particularly distinctive is that the edge of the blade is flat or gently wavy, while the centre is all puckered up. If you allow a length of sugar kelp to dry out, a white powder forms on the surface, which is deliciously sweet with a hint of the sea. Although, if you are going to start licking bits of seaweed, I suggest you consult a proper identification guide first. While sugar kelp is popular in places such as Japan, other nations are not so keen.

Instead you could turn to glycyrrhizin, found in the woody roots of *Glycyrrhiza glabra*, more commonly called the liquorice plant, and used in the production of liquorice sweets. While glycyrrhizin is only fifty times sweeter than sucrose, it does seem to linger on your taste buds, giving liquorice one of its unique characteristics. It is also best eaten in moderation, as not only can it cause a rise in blood pressure, but it also has laxative effects.

Finally, my last example of an artificial sweetener from a natural source is stevia, or more precisely the steviol glycoside group of chemicals derived from the South American sugarleaf herb. These chemicals are about 150 times sweeter than sucrose, temperature stable, acid-resistant and non-fermentable

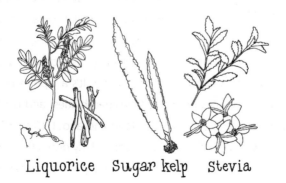

Liquorice   Sugar kelp   Stevia

by yeasts. All of which has made them very popular as food additives, to the extent that both the The Coca-Cola Company and PepsiCo have produced stevia-based sweeteners.

What these sugar-free sugar substitutes have in common is that they all bear some structural resemblance to sucrose itself. It therefore comes as no surprise that our taste buds detect them as sweet, as they all possess the key to the sweetness lock. So how, then, does the super-sweet lugduname work? There are a number of theories about our ability to detect sweetness, and the most recent is called the multi-point attachment theory, developed by biologists at the University of Lyon in France. In this theory, the sweetness receptor on the tongue detects not one big structural region but up to eight, smaller and spaced-apart areas – it looks like a molecule doesn't need to contain all eight regions to register as sweet. It's not so much a lock and key model as a sack full of locks and a key ring crammed with tiny keys. This also gives us an elegant way to envisage why super-sweet lugduname doesn't look like sucrose. While the molecules are dissimilar, they must each open enough of the eight locks to qualify as sweet. It may be that the sub-set of locks on the sweetness receptor they each open is different, but our tongue is clearly a lot less discriminating than we would imagine, and all sugar is not equal.

Light and fluffy cake chemicals

I would argue that there are few things more pleasing than a deliciously fresh and spongy cake accompanied by a hot cup of tea. The production of such an appetizing, light and fluffy example of the baker's art is surprisingly simple. You need two basic things: something that produces lots of bubbles of gas, and a way to trap those bubbles inside the yummy cake. The second part of the job is almost universally achieved by adding egg to the mixture, but there are a couple of options for creating the bubbles.

It is perfectly possible to create your bubbles by whisking air into the eggs, but far easier and more reliable is to use a bit of clever chemistry instead. Most cakes I bake rely on self-raising flour to produce the requisite bubbles of gas. The self-raising part of the flour comes from the inclusion of baking powder at 5 g to every 100 g plain white flour (1 oz baking powder to 1¼ lb flour). What, then, makes baking powder so good at blowing bubbles?

There are two key ingredients in baking powder: the first is sodium hydrogen carbonate. This is its official IUPAC (International Union of Pure and Applied Chemistry) name, but it is more commonly known by a number of colloquial names that include: bicarbonate of soda, baking soda, cooking soda or just plain bicarb. The soda bit of the name refers to sodium and is not the interesting bit. What makes this chemical so useful to us is the hydrogen carbonate as, when this is dissolved in anything acidic, it turns into carbonic acid, which quickly breaks down into water and carbon dioxide gas. All of the bubbles inside a cake are made using the breakdown of hydrogen carbonate and are filled with carbon dioxide gas.

As an aside, sodium hydrogen carbonate will also break down to produce carbon dioxide if you heat it above about 50 °C (122 °F). Some baking powder is labelled as 'double-acting', and this relies on not only the acid reaction but also the heat breakdown of hydrogen carbonate to produce bubbles.

What makes baking powder really cunning is the inclusion of a second key ingredient, along with the hydrogen carbonate, called disodium dihydrogen phosphate, which sounds terribly complicated. However, this second component is just a dry powdered acid. When you mix it in water, disodium dihydrogen phosphate creates a slightly acidic solution, equivalent to adding a splash of lemon juice or vinegar, but without the smell.

Both elements of baking powder are completely inert when left mixed together in your cupboard but dissolve them in water, or anything containing water, such as milk or eggs, and the chemistry starts to happen. The dihydrogen phosphate makes your cake mixture a little bit acidic and the hydrogen carbonate then starts to immediately produce carbon dioxide gas. Which is why, once you have added liquid to your baking powder or self-raising flour, you want to get your skates on and get the mixture in the oven, so you can set the egg and trap the bubbles. If you leave cake mixture on the kitchen counter for too long before baking it, your cake will not be as light or fluffy, although it may still be delicious.

Putting the cracker into prawn

Whether you call them prawn chips, shrimp crackers or prawn crackers, when they arrive on the table at Asian restaurants, these crispy, curled morsels always disappear before you have time to reflect on what they are, how they are made and if they actually contain any prawn or shrimp. It often comes as a surprise to people that prawn crackers do indeed contain prawn, and the ingredients list is unusually short for what is clearly a heavily processed food. The prawn content is somewhere around 10 to 15 per cent, but the bulk of the cracker is made up of tapioca starch. Converting simple starch with prawn to a bubbly crisp is a marvel of food manufacture. The first step is to create a disc of starch plastic.

The key ingredient in prawn crackers is tapioca starch, which itself contains some scientific surprises. It comes from the tuberous roots of the cassava plant. These roots look a bit like sweet potatoes and are a staple crop for large populations of people living in the tropics. Given this, it's a little alarming to discover that cassava is also a rich source of cyanide and can lead to both acute and chronic poisoning. There are different varieties of cassava, and the ones known as bitter cassava contain dangerous amounts of a chemical called linamarin, which is basically glucose attached to cyanide. When the root is peeled, or chopped up, enzymes are released that break down the linamarin and release the cyanide. With this in mind, a vital part of the tapioca starch extraction process is to remove all of the cyanide. Step 1 is to finely grate the cassava, which starts the breakdown of linamarin and the production of cyanide. In step 2, you chuck it all in a vat of water and leave it to soak for a couple of days. The cyanide will dissolve into the water, and by changing the water a few times to give the grated cassava a rinse, all of the cyanide is flushed away. You then take the resulting mush and squeeze all the liquid out. The mush you chuck away and the milky white liquid you leave to evaporate. What remains is a very fine, and very pure, starch powder.

It is possible to make prawn crackers by hand, the artisanal way. Take minced prawns, mix with tapioca starch and add a little water to make a dough. Then form the dough into a sausage and steam it. This produces an extremely unappetizing, gluey blob of cooked starch that needs to be dried for a couple of days. It is then cut into thin slices and further dried for a few more days. You end up with a disc of what looks and feels like plastic. It will be slightly translucent and, depending on how thin you managed to make it, pretty tough stuff. Finally, drop this into hot oil and it magically expands, puffing up and turning into a prawn cracker.

If you think this sounds like an involved process, the industrial method is possibly even more extreme. The key to making prawn crackers is to get the moisture content just right. For bulk manufacture they use tapioca starch, dried prawn powder and a tiny amount of water. The resulting powder is placed in a machine that compresses and heats the starch mixture. The pressure involved is huge, at up to 2 tonnes per square centimetre (28,500 pounds per square inch). At this point the starch melts, flows together and turns into a thermosetting plastic, which is just a fancy way of saying it's a solid material that goes soft and runny when you heat it. This molten plastic starch is pooped out of the machine in little flat, translucent white discs. You can buy these uncooked prawn cracker discs in Chinese supermarkets and Asian food shops. In this form, as long as they are kept in a sealed plastic bag, they have an enormously long shelf life.

I said the water content at the start of the process was critical and when you drop your plastic starch disc into hot oil it is vital you have the right amount of water trapped inside the starch plastic. Two things happen: first, the thermosetting plastic starch heats up and turns soft and runny again. Secondly, the tiny amount of water it contains vaporizes, turning to steam and expanding in the process. As the specks of water turn into bubbles of water vapour surrounded by shells of plastic starch, the flat disc puffs up and turns into the familiar bubbly crisp. You then whip it out of the oil, before the starch has a chance to start browning and, as it cools, the starch plastic goes back to being a hard and brittle substance. But now, rather than a solid disc we have an aerated mass of crispy, crunchy loveliness. This may seem like a niche cooking method, but exactly the same process is used to make a number of everyday food products from puffed breakfast cereals to popcorn and even those, admittedly inedible, puffed packing worms you find crammed into boxes to protect the contents in transit. Not so unusual after all.

Egg white, not see through

Consider this: when you cook an egg, be it chicken, duck or quail, the egg white, or albumen to give it its correct name, turns from completely clear liquid to a solid, translucent white. On the other hand, the yolk remains the same colour, although it also changes consistency. Why should the transparency of one change, but not the other?

The egg of a bird such as a chicken is packed with the protein, fat and minerals needed to make a chick. The yolk contains the majority of the calories in an egg, and is the primary source of nutrition for the developing embryo. It's the bit that has all the fat, unlike the albumen of the egg that is almost pure protein mixed in water. The albumen is there to support and protect the yolk although, eventually, it too is used up in the process of creating a baby chicken. The proteins within the uncooked albumen of the egg are called albumins, and are made up of long chains of hundreds of amino acids. Along the length of these chains are charged chemical groups that will stick to other charged groups along the same chain. Consequently, the proteins roll themselves into tiny little balls as all the charges pair up, and glue it together. The albumen of an egg is a solution of these albumin protein molecules floating about in water.

You now need to get your head around what makes something transparent or the opposite, opaque. On a molecular level, uncooked egg white is packed with molecules of water and protein, each of which is made up of constituent atoms.

At this scale it seems unlikely that light could penetrate far, let alone pass through. However, go beyond the scale of the atom and into the realm of subatomic particles and all this changes. All atoms are made up of a central nucleus, surrounded by a cloud of orbiting electrons and the nucleus takes up a tiny portion of the space inside the atom. There are numerous popular analogies to illustrate this involving sports stadia and peas, but the core concept is that inside any atom there is very, very little of anything – it is mostly just empty space filled with a cloud of electrons.

When a ray of visible light hits an atom it is almost certainly not going to hit the nucleus, but will pass through the cloud of electrons. Since we are now talking about things on the sub-atomic scale, we have entered the realms of quantum effects. Electrons can only exist in certain predefined energy levels. The reason behind this, without delving into too much quantum weirdness, is analogous to electrons possessing several resonant frequencies (see page 103). The possible energy levels depend on the type of atom and to what else it is joined.

A ray of light has a particular amount of energy associated with it, defined by the wavelength, or colour, of the light. When light passes through electrons, they can absorb this energy and jump to one of their higher energy levels, but only if it's exactly the correct amount of energy. An electron can't jump halfway to a new energy level, or overshoot a bit. The energy has to be just right. It turns out that in egg white, full of water and proteins, all of the electrons have energy levels that are spaced out too far. When visible light hits the egg

THE SCIENCE OF EVERYDAY LIFE

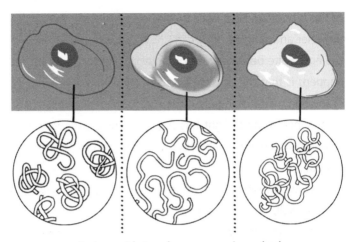

Amino acids tangle as an egg is cooked

white, it has the wrong energy to be absorbed by the electrons. Since it is not being absorbed, it passes straight through and the egg white liquid appears transparent to light. It should be noted that water and raw egg albumen is not transparent to higher energy, ultraviolet light. This kind of light does have the right amount of energy for the electrons and is consequently absorbed.

All of this changes when you start to heat the egg white. At about 60 °C (140 °F), the first albumin proteins begin to change their structure. By the time you hit 80 °C (176 °F), there is a mass breakdown of order within the egg white. The curled-up balls of amino acids that make up the albumins are shaken so violently by the heat that the chemical bonds holding them together as little balls start to come apart. The balls unravel and our egg white is filled with long chains of amino acids which become entangled and stick to each other. The up-shot of this is twofold. First, since the protein molecules are all stuck together and tangled, they can't now freely move about, and the egg white becomes a wobbly solid. The second thing that happens is that the possible energy levels within the electrons in the egg white change, so that they can absorb visible light. Now, when a ray of light hits the egg white, it doesn't pass through, its energy is absorbed and the egg white appears opaque.

It's worth wondering what happens to all this absorbed energy. Well, it's released by the electrons, as they sink back to their lower energy levels, in the form of light. However, it's released in all directions, not necessarily the direction the original ray of light was travelling. While some of it will carry on into the egg white, at least half will be reflected back towards the original light source. All of which makes the egg opaque and white.

So, now that the transparency of egg white becomes clear – pun intended – what of egg yolk, why is this not clear? In this case it's a bit less complicated than it having the wrong electron energy levels. Egg yolk is not just water with protein dissolved in it; it's also full of tiny blobs of fat. When light hits these it reflects off the surface of the blobs, scattering the light.

Given how many things need to be just right for a substance to be transparent, it's remarkable that anything is. Don't even get me started on how a solid, such as glass, manages this trick.

Smoked but not cooked

A while ago I was asked by a young person if smoked salmon was cooked. Now, this is the sort of question that elicits in parents, among whom I count myself, a spasm of worry, and a response along the lines of *'Don't worry about that now, just eat it and don't make a fuss.'* At the time, the young person in question was one in whom the idea of not accepting half-baked answers had been carefully inculcated. I had only myself to blame, and was forced to think somewhat harder about a more satisfying response.

Ultimately, the answer to this question depends not on the salmon but on what you mean by cooked. The dictionary definition of the word is the preparation of food with the use of heat, but significantly it does not specify how much heat and with what result. A less semantic and more scientific definition would include something about both food preservation and the unravelling of the proteins in the food.

If you are trying to achieve preservation and unravelled protein, then heat is the simplest way forward. Temperatures above 70 °C (160 °F) are needed to kill bacteria, and this is also the temperature, not coincidentally, that protein molecules start to come apart. Above this temperature, the long chain molecules of proteins uncoil in a process called denaturation. This usually leads to a distinct change in appearance (see *Egg white, not see through* on page 22), and in the case of salmon is characterized by a change from dark, translucent pink to much lighter, opaque pink.

Since this colour change does not happen in smoked salmon that must mean it is not cooked. Right? Well, sort of. The trouble is that heating is not the only way to preserve food. Traditional smoked salmon, in the United Kingdom, is made using a two-step process. First, the raw fish is covered in salt and left for 24 hours. This sucks a load of water out of the fish, drying it out by about 10 per cent of its initial weight. The salt that permeates the flesh of the fish that sucks the water out also kills most of the bacteria that may be lurking inside it. Following this, the fish is hung up in a smoky room for about 12 hours at a temperature of no more than 30 °C (86 °F). The smoke itself does little beyond providing flavour, although there is evidence that some of the chemicals in the smoke may be antibacterial. What the smoke does do is dry the surface of the fish out by yet another 10 per cent. This drying, combined with smoke and the presence of salt, makes the surface of the fish decidedly inhospitable for bacteria, and the fish has been preserved, to a small degree.

Where does this leave us with the original child's question? Smoked salmon is prepared by the application of gentle heat, and the proteins in the fish are not denatured, but it is preserved. By our dictionary definition this merits being labelled as cooked, but by the more rigorous scientific definition of denaturation *and* preservation, it is not cooked. So, finally, my attempt at a less half-baked answer is that smoked salmon is half-cooked.

A cold loaf is a stale loaf

Refrigeration has changed the way we eat and the way we farm. By placing food at a few degrees above freezing you can slow down the growth of bacteria and moulds and increase the shelf life of dozens of foods, from yoghurt to whole chickens. The low temperature also helps to keep food moist by reducing evaporation, and in the case of some fruits it will actively stop them from ripening, should you so wish. Taken together, the refrigerator in the kitchen and refrigerated transport around the globe has allowed us to enjoy the unseasonal variety that we have become accustomed to in our supermarkets. However, you should never refrigerate bread. Freezing bread is fine, but never, I repeat, *never* keep it in the refrigerator.

Bread can include all manner of ingredients, but at its heart are just three: flour, water and yeast. The yeast is there, as a living microorganism, to grow and produce carbon dioxide gas bubbles for a light, fluffy texture in the baked loaf. It is the flour and water that play into my advice on bread refrigeration.

It should not come as news, I hope, that flour is made by grinding the seeds of the wheat grass. These seeds are made up of three parts. The outside coat of the seed, or the bran, is rich in fibre, but not much else. Inside this is the germ, that when the seed is planted becomes a new wheat seedling. Finally, and filling up most of the space inside the bran, is a big blob of starch mixed with a bit of protein. Whole wheat or brown flour contains all three parts, but white flour is just the ground-up starchy blob with protein. If you mix white flour with water and squidge it around a bit you will make a springy dough rather than a gluey mess. It's the protein in bread flour, called gluten, that puts the spring into bread, but it has no bearing on refrigeration, so I won't be mentioning it again.

What *does* make a difference is the form the starch takes in the wheat seeds before it is ground up to make flour. Starch, in plants, is made by hooking together long chains of a type of sugar called glucose and then allowing these chains to stick to each other. The plant forms tiny granules of starch as a store of food in the seed and, since the starch is neatly lined up in the granules, it is described as a crystalline structure. When you take these granules of starch that make up the bulk of flour and mix in water, the water wriggles in between the long glucose chains. This splits up the neat crystalline structure, the starch granules swell, and they become softer and gelatinous. You can see this if you pour boiling water onto cornstarch – it immediately turns into a sticky goo. While this does not sound very appetizing, it is this starch gloop that makes bread soft and moist. So, all well and good, you have made a lovely soft loaf of bread, full of gelatinous starch.

If you leave the bread out on a table it will begin to go stale, partly because the water in it gradually evaporates, but also because the starch slowly switches back to its crystalline form. This second process is called retrograding and when it happens, water is squeezed out of the gelatinous starch, and even though the water may still be in the bread, it tastes drier and becomes stale. The crucial point here is that the process of retrograding is dramatically increased at temperatures between −8 °C (18 °F) and +8 °C (46 °F). If you put bread in the refrigerator, at 5 °C (41 °F) the starch will retrograde and the bread will go stale. Even if the bread is wrapped tightly in plastic to prevent it drying out by water evaporation, refrigerated bread will go stale more quickly than bread at room temperature. The bread will taste dry, even though the water content will hardly have changed.

It's not all bad though, since starch does not retrograde much below −8 °C (18 °F). It is perfectly sensible, if you want to prolong the life of a loaf, to freeze bread at about −20 °C (−4 °F). You can also often recover bread that has gone stale in the refrigerator by warming it gently, especially if the bread has not lost any of the water squeezed out of the starch. Just pop the bread in a moderate oven for five minutes and it will not only become crusty but also taste fresh. Of course, bread kept at room temperature will go mouldy much quicker than refrigerated bread, so it's your choice: mouldy or stale.

Spicy stuff

Lurking within the average spice collection in your average kitchen is a veritable wealth of interesting and naturally occurring drugs. To a pharmacist, some of these drugs share a very specialized biochemical action and, to those of us with a culinary mindset, they all impart a piquant spiciness to things we eat.

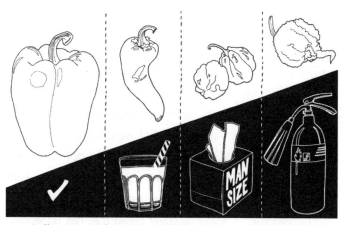

Bell pepper, jalapeño, Scotch bonnet, Carolina Reaper

The best known of these drugs is called capsaicin and it is found in chilli peppers of all shapes and sizes. The relative heat of chilli peppers depends on how much capsaicin they each contain, and this heat can be measured and quantified using a system invented by Wilbur Scoville in 1912: the eponymous Scoville scale. On this scale, a bell pepper ranks zero, jalapeño peppers about 2,500 and a Scotch bonnet chilli comes in at between 100,000 and 350,000 units. However, these chillies are babies compared to the super-hot chilli varieties. The current world record holder, the Carolina Reaper, an ugly, bright red, wrinkled chilli, has clocked well over 2,000,000 Scoville units. Unfortunately, this scoring system is rather untrustworthy, as it relies on a panel of five tasters sampling diluted extracts of the chilli under test. When three out of the five testers agree that they can just detect heat in their mouth, that dilution sets the Scoville value. Depending on the people doing the tasting, the results can vary wildly. It's believed, based on a different type of chemical test, that pure capsaicin would have a Scoville rating of sixteen million, which is surely mind-blowingly hot enough for even the most ardent chilli fan.

We describe the effect of chilli and the capsaicin it contains as 'hot' for a good reason. On your tongue and in the lining of your mouth are nerve-cell endings that detect high temperatures. If you take a mouthful of soup that is too hot, you know instantly, because these nerve endings detect the heat that could cause damage to your mouth. The very tips at the end of the nerve cells have cell membranes that contain a protein that is triggered into action at 43 °C (109 °F). At this temperature and above, the protein kicks into life and opens a hole through the nerve-cell membrane, allowing calcium ions to flood in. This in turn starts a nerve impulse that shoots off to your brain, where you register it as heat and pain. This protein has the extremely unmemorable name of 'transient receptor potential cation channel subfamily V member 1', or TRPV1 to its friends. It turns out that not only will heat activate TRPV1, but capsaicin has exactly the same effect when it sticks to the protein. So, the sensation of chilli in your mouth is *exactly* the same as something hot, for the reason that both things are detected by the same nerve cells.

However, chilli pepper is not the only hot thing in your cupboard. Black pepper, ginger, mustard and even Sichuan peppercorns all create heat in your mouth, but each contains an entirely different chemical to do so. What links all of these taste sensations is that the effect all comes back to our friend, the protein TRPV1. Each hot spice contains a compound that will activate our heat-detecting nerves. Black pepper contains a compound called piperine, with 100,000 Scoville units, and ginger has gingerol at just 60,000 Scoville units. The mustard family, that also includes horseradish and wasabi, contains a somewhat different chemical called allyl isothiocyanate. While this also triggers our heat-sensing nerves, it is much more volatile, meaning it can turn into a gas very easily. So, when you take a spoonful of mustard, the allyl isothiocyanate turns into a gas and is whisked up your nose where it activates the heat-detecting nerves there, which in turn makes you cry and usually clears your sinuses.

The last spice on my hot list, the Sichuan peppercorn, has even more tricks up its sleeve. This spice is found in Asian cooking and is one of the components of Chinese five-spice powder. It's harvested as the skin of tiny berries from a relative of the citrus family. The active chemical it contains, hydroxy-alpha-sanshool, has the familiar hot taste of the capsaicin but also a strange numbing or tingling effect on the mouth. The heat we feel is our old friend TRPV1 turning on, but the science behind what causes the numbing is still not completely certain. It looks as though the hydroxy-alpha-sanshool might work on another set of proteins in nerve cells that are responsible for our sensation of touch.

Ironically, all of the plants that produce these spices that many of us crave in our food have evolved to contain these active chemical ingredients specifically to dissuade animals from eating them. It is only the perversity of humans who delight in mimicking the pain caused by heat that drives us to dose our food with them so liberally.

Taking the biscuit and the cake

When is a biscuit a cake and when is a cake a biscuit? To understand this question, we first need to clear a few things up. In this particular instance, the biscuit I am referring to is the British, Irish and European version of this tasty product, which on the other side of the Atlantic would be called a cookie. If you sit down to eat a biscuit in the United States of America or Canada, you will probably be served what the British would call a savoury scone, or maybe a cobbler. If you are reading in Australia, I'm afraid you will have to work out what I'm on about from context – I have no idea what is the Aussie term for biscuit. Good luck.

Diversions aside, there is a simple way to determine what kind of bakery product you have: just leave it unwrapped on the counter top for a day or two. If it's a cake, it will dry out and become hard and brittle, whereas if it's a biscuit it will become soft. The crux of this deceptively simple test is the in-depth science of what holds together a cake or a biscuit.

The structure of a cake is made of a matrix of flour, held together with egg around bubbles of gas (see page 17 and page 18). When you bake cake mixture, you heat the egg and it turns from a liquid to a flexible solid as the eggy protein molecules unravel and become tangled up. It is this flexible but solid egg that gives cake its softness. The flexibility and bendiness of solidified egg that makes a cake spongy is due to the presence of water molecules. These molecules allow the unravelled egg protein just enough interaction that they turn into a solid, but not so much that they set rock hard. As the

34

water evaporates from a cake, the eggy matrix binds together ever more rigidly, and the cake will become hard and brittle.

Biscuits, on the other hand, are held together with fat and sugar. Clearly this is a generalization as there are biscuit recipes that include egg, but they also include a high proportion of sugar. When the grains of sugar in a biscuit dough are oven-baked, they melt and flow together. Biscuits straight out of the oven are soft and bendy, but as they cool the molten sugar crystallizes and turns hard. If you protect your biscuits from the water in the air by placing them in an airtight container, all will be well. However, when exposed to even a small amount of water vapour, the crystallized sugar will pull water out of the air and become partially dissolved. When this happens the sugar loses its rigidity, the biscuit loses its crunch and it goes soft.

You may feel that being armed with the true definition of a cake and a biscuit is of trivial importance. However, the semantic difference between a biscuit and a cake is a multi-million pound question. In the United Kingdom you can buy a packet of Jaffa Cakes at any supermarket or convenience store. For the uninitiated, a Jaffa Cake is a 64-mm (2½-in) disc of Genoese sponge, topped by a smaller disk of orange-flavoured jelly that is then slathered in dark chocolate. Taken together, you have a deliciously moreish morsel of loveliness, to the point that a packet of Jaffa Cakes is a binary object, by which I mean it has only two states: unopened or empty. There have been rumoured sightings of half packets, but the evidence is debatable. The big and valuable question, though, is this: are Jaffa Cakes a biscuit or a cake?

In the United Kingdom, VAT, or Value Added Tax (sales tax), is not charged on the sale of cakes, nor is it charged on biscuits unless the biscuit is covered in chocolate. Why and how they concocted this rule is way beyond the scope of this book. If the Jaffa Cake is defined as a cake, it does not have VAT added, but if it is a biscuit, because it is chocolate-covered it must carry VAT. The problem is that Jaffa Cakes come in packs like biscuits, are eaten like biscuits and are stacked on supermarket shelves next to biscuits. On the other hand, they are made from a bit of sponge cake topped in chocolate with a 'smashing orangey bit'. Fortunately for lovers of Jaffa Cakes, a ruling at the United Kingdom VAT Tribunal in 1991 decided that a Jaffa Cake was and always will be a cake. The clinching evidence was when the manufacturers showed that a Jaffa Cake when left out in the air will go hard. I find this to be a remarkable piece of research, as whenever I find myself near a Jaffa Cake that has been left out, it disappears.

The bottle, the wine and the oxygen

Connoisseurs have raised the classification of the various flavours in wine to an art form. Yet despite all the subtleties, the simple act of opening a bottle can have a profound effect on its taste. Wine has a complicated relationship with oxygen;

it's an essential component during some stages of production, but once the wine gets to the consumer, oxygen is a bad thing. The most common way to encounter wine that has had a disastrous encounter with oxygen, or has been oxidized, is when you half drink a bottle of wine and then leave it for a few days. On your return, the wine will have lost its fruity smell and seem flat and boring. White wines are particularly prone to this and will turn an amber colour, take on the smell of sherry and, in bad cases, your nose will be prickled by a hint of vinegar. In red wines it takes longer to happen, but when it does the wine changes from a vivid purple to a more muted brown, takes on a bitter taste and will eventually also turn vinegary.

All of these changes start the moment you open the bottle of wine. Up until that point the bottle was essentially airtight, containing only a tiny bubble of air, and an even tinier bit of oxygen. The very act of opening the wine and pouring a glass allows sufficient oxygen to come into contact with the wine to begin oxidation. A ticking clock has been started and your wine will inexorably begin to change.

The oxygen reacts primarily with a class of compounds found in the wine called phenols. These phenols come in different varieties that can be grouped as complex or simple. For complex phenols, especially those in red wines, this oxidation turns out to be a good thing as it mellows their astringent taste. Which is why it's recommended that you allow red wine to breathe for a little while, oxidizing some of the phenols to give the wine a richer, more mellow taste. This is also why red wine is less prone to being spoilt by oxidation, as its complex phenols tend to mop up the oxygen. With simple, or monomeric, phenols, the oxidation is not so beneficial as it produces hydrogen peroxide, an extremely reactive chemical. This attacks the alcohol in the wine, converting it first to acetaldehyde, the smell of sherry, and then to acetic acid, which is the chemical name for vinegar.

It should be noted here that all I've so far described is the fault of oxidation, which is commonly mistaken for, but completely different to, a wine being corked. When a wine is corked it contains minuscule quantities of a chemical called trichloranisole. This gives the wine a vile smell that's described as mouldy, or like two-week-old sweaty sports socks. Either way, it's not a pleasant aroma, and very obvious when present. The trichloranisole leaches into the wine from the cork, where it was originally produced by fungi, when the cork was part of the bark of a cork oak tree. I should also point out that if wine has bits of cork floating in it, this does not mean it's corked – just that the corkscrew used to open the bottle has made a mess of things.

It's not possible to completely halt the process of oxidation in wine once the bottle is open, but it should be possible to slow it down. There are a variety of cunning devices available to the wine drinker that claim to help with this problem. The most common system uses a rubber stopper and a small hand-powered vacuum pump to remove the air from inside the bottle. No air inside the bottle means no oxygen, which means no oxidation. Which is great, except that if you get too enthusiastic with the vacuum pump you will also draw dissolved gases out of the wine, and this can have an effect on the taste. Another system has you pour the remaining wine into a mini bottle or a plastic pouch. This method relies on creating a smaller volume of air and consequently less oxygen above the leftover wine. But some experts will tell you that the very act of decanting wine from one bottle to another is enough to introduce sufficient oxygen that the wine will spoil.

It turns out that the most traditional method is probably also the most reliable: pop a cork back in the bottle and store it in a refrigerator. The cork will stop any more oxygen getting into the bottle and all chemical processes slow down at lower temperatures. A white wine will last four or five days before oxidation, and reds should be good for over a week. Of course, this method, while fine for white wine served cold, is less convenient for most reds.

Wine is an incredibly complicated blend of flavour molecules. The unique aroma of each different wine is created by some fiendishly fiddly chemistry that takes place in a predominantly oxygen-free environment. So, it should come as no surprise that adding even limited amounts of a highly reactive molecule such as oxygen can be disastrous. But there is one sure-fire, guaranteed way to ensure your bottle of wine doesn't become oxidized: never, ever leave a bottle of wine part-finished.

The tearful subject of the onion

The onion is one of our oldest cultivated vegetables. The Romans and the Ancient Greeks wrote about it, and 5,000 years ago the Egyptians used onion seeds in their mummification rituals and painted pictures of onions on their tomb walls. It can be traced even further back than this to the early Bronze Age – onions have been discovered in remains 7,000 years old in Palestine. You would think that with millennia of cultivation we would have got to grips with the basic problem of onions: they make you cry.

Take a knife and start to slice up the flesh of an onion. As you do this, you break open lots of the unusually large cells of the onion. Within these cells are two chemicals that normally don't come into contact, since they are contained in different cellular compartments. By cutting open the cells you also break these compartments and the chemicals mix. The first of these substances is a group of protein-building blocks called amino acids, linked to a sulphur and oxygen atom. When these sulphur-linked amino acids encounter an enzyme known as alliinase, they produce a highly reactive sulphenic acid. (And I have spelt the word alliinase correctly. It comes from allium, the scientific name for the onion genus of plants, and for reasons unknown the enzyme has an extra 'i' thrown in for good measure.)

The creation of the sulphenic acid is not the end of the chemistry; a second enzyme gets involved. The grandly named lachrymatory factor synthase gets to work on the sulphenic acid and produces – you guessed it – lachrymatory factor, or syn-propanethial-S-oxide – I think it would be wise to stick with lachrymatory factor in this instance. Now we are getting to the tearful end of the story, as lachrymatory factor is a highly volatile liquid that turns into a gas that floats up to your eyes.

It's possibly surprising that the see-through part of the front of your eye, the cornea, is packed with sensory nerve endings. These are there to detect anything that touches the delicate cornea, and when this happens we unconsciously blink and also produce tears to flush the irritant away. The lachrymatory factor sticks to these nerve endings, fooling them into believing that something hot has touched our cornea. We feel this as a burning pain, even though there is no heat there, and we begin to cry, or to lachrymate, to use the fancy word. There are many chemicals that can cause the same reaction – capsaicin, for example (see page 31), but it is only onions and their relatives that produce a gas that does this.

While this explains what happens, it sheds no light on why the onion has this convoluted chain of chemical events waiting inside it. For this we need to turn to botany, and an understanding of herbivory. The onion plant is biennial, meaning that it lives for two years. In its first year it grows from a seed into a fan of thick, but hollow green leaves. Along the way, it creates a food store for itself in the base of these leaves, and it's these swollen leaves that make up the onion bulb. This bulb overwinters and in the spring it sends up a shoot, more leaves and a flower spike. The flowers in turn set seed and the whole cycle starts again. Clearly, from the onion plant's perspective it is crucial that the onion bulb, chock-full of stored energy, stays unharmed in the ground over the winter. To this end, the onion has evolved a sequence of unpleasant chemicals. If an animal chomps on an onion bulb, the lachrymatory factor is released, the eyes of the herbivore begin to burn and they sensibly leave it well alone.

Unfortunately for onions, some of the lachrymatory factor remains in the onion and breaks down to create delicious flavours. Humans are perverse enough to endure the pain just for a taste of onion.

We are so keen on onions that there is a huge mythology surrounding ways to avoid the tears. These methods range from the bizarre and useless, like biting on a wooden spoon while chopping the onion, to the deeply inconvenient, such as cutting onions under running water. There are, however, a few soundly scientific solutions. Since the lachrymatory factor must contact your eyes, wearing a set of swimming goggles completely stops any tears. It does make you look rather silly, so if that's not for you try opening a window or turning on a fan to create a breeze and blow the lachrymatory factor away. The solution used by chefs who are in the business of chopping onions is simpler still: chop the onion quickly. It takes about 30 seconds before the chemical reactions start producing lachrymatory factor. With a really sharp knife and the proper cheffy technique for chopping onion, it will take you less than 30 seconds to get the job done. I should add that you also need to get the chopped onions into a pan, coated in oil and cooking immediately. It's no good chopping an onion really quickly and then leaving it on the board, merrily producing lachrymatory factor galore.

There is one further scientific solution to this tearful problem. In 2008, Colin Eady and his team of biologists working in New Zealand found a way to genetically modify onions so that they could shut down production of the lachrymatory factor synthase enzyme – no enzyme, no lachrymatory factor, no tears. In addition they claimed that by not producing lachrymatory factor, all the tasty chemicals stayed within the onions, making them even more delicious. Since these are only initial results, it will be many more years before truly tear-free onions show up on the supermarket shelves. Until then, carry on wearing goggles, and learn to chop faster, in a breeze.

The Heart of the Home and Kitchen Science

The invention that changed our food

Refrigeration is not just about keeping your beverage of choice cold; it's the cornerstone of our Western food culture. Take something essential, such as bags of ready prepared salad leaves. I for one would never eat anything more exciting than limp leaves of floppy lettuce without those marvellous bags of greenery. If you leave a salad bag out of the refrigerator, within just a few days it turns into the most repulsive black slime. It is a fascinating and quite magical transformation, but not the issue. By refrigerating mixed mizuna and rocket leaves, you slow the process of cellular breakdown and bacterial growth. Without refrigeration it would not be possible to harvest, wash, bag up and transport the salad before the black slime kicked in. It is, of course, not just for bagged salad that we need refrigeration. Without it, half the shelves of supermarkets would be empty. The most popular fruit in both the United Kingdom and the USA is the banana. Yet without refrigerated transport at 13 °C (55 °F), bananas, grown in the tropics, would turn into overripe black mush while they were still weeks away from our shores.

The idea of using cold temperatures to extend the shelf life of food has been known for centuries. The great Francis Bacon, the early seventeenth-century polymath rather than the twentieth-century painter, is generally credited with inventing the frozen chicken. It was not all he did, but certainly the only thing pertinent to refrigeration. In the early spring of 1626, while en route to Highgate in North London, for reasons unrecorded, Bacon decided to buy an eviscerated chicken and stuff it with snow, thus demonstrating that refrigeration was a remarkable way of keeping food fresh for longer. Unfortunately, as the whole escapade was an off-the-cuff experiment, Bacon must have been unsuitably attired for the snow. He caught a chill, which became pneumonia, and he died shortly thereafter while still at Highgate. He died a martyr to his science and sadly history does not record the fate of the world's first oven-ready frozen chicken.

The cold that your refrigerator produces is the result of a simple bit of science, namely evaporative cooling. Next time you hop out of the shower, pause for a moment to consider why it is so very cold. After all, when you stripped down to get into the shower it did not feel that cold. So why when you get out of the shower does it feel colder? It is not the room that is colder, but the water that is evaporating off your skin that is cooling you. It is an effect that was first demonstrated in 1756 by a canny Scotsman called William Cullen, at a public lecture on the subject of how to make things cold. This took place in Edinburgh, Scotland, so the audience were presumably well acquainted with the subject of being cold. In the lecture he showed that if you allow a liquid called di-ethyl ether to evaporate, it cools so much that it can be used to freeze water, making ice. The reason he used di-ethyl ether is because it has a very low boiling point of 35 °C (95 °F), less than body temperature of 37 °C (98.6 °F). William's talk went well by all accounts, but the ice he made failed to ignite any inventive sparks and it took 150 years until the principle became the machine.

Inside your home refrigerator, exactly the same principle is at play. William's table-top demonstration is made practical for us all. In a series of pipes, a special liquid known as a refrigerant, alternately evaporates and cools, then condenses and warms. Your refrigerant does the same thing as William Cullen's di-ethyl ether. The place where evaporation happens is the vertical plate at the back of the inside of your refrigerator. The refrigerant turns to a gas in the tubes in this plate by sucking in heat from the inside of the refrigerator for energy, which makes the refrigerator colder. The now gaseous refrigerant makes its way to the metal grill at the back of the

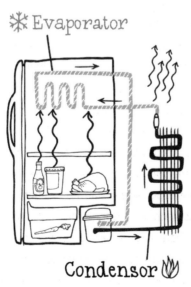

✳ Evaporator

Condensor 🔥

refrigerator. I'm sure you know the one I mean; it probably has a thick layer of dust over it, catches things that fall down the back of the refrigerator and, most importantly, gets warm to the touch. Inside these tubes the reverse process takes place. The refrigerant turns from a gas back into a liquid, and releases energy. This energy comes out as heat, which is dispersed to the air behind your refrigerator. To get this system working, all you need to do is pump the refrigerant around the tubes. By connecting up tubes of different diameters, you create areas of high and low pressure within the sealed system and the refrigerant can be conveniently made to evaporate and condense in all the right places. While the layout of tubes and the refrigerants we use have changed, the principle behind the science has not. It's a remarkably simple system that has remained mostly unaltered for over a hundred years.

So, is it possible to build a better refrigerator? There are other technologies now that allow for cooling, such as the Peltier effect that directly converts electricity into a difference in temperature, but it's not very efficient and can only be used to make very small refrigerators. If you want to make your refrigerator more efficient there are a couple of things you can do. First, don't open it, which I grant you is not the most practical advice, but each time you do, the cold air literally falls out of the refrigerator and is replaced by warm air. What you can do, more realistically, is keep your refrigerator full. Then, when you open the door there will be less air inside to fall out and the inside of the refrigerator will stay colder. Finally, for the really keen, you could insulate your refrigerator using polystyrene sheets. It will help keep heat from seeping into the refrigerator and halve the energy that your refrigerator uses. A word of caution: if you attempt this, make sure you don't cover up the warm coils of pipe at the back of the refrigerator, or they won't be able to do their job of dissipating heat.

While it may not be the most glamorous of inventions, or the most spectacular to look at, there is no doubt in my mind that the refrigerator and its super-charged cousin the freezer have had the greatest impact, of any invention, on the diet of the Western world.

The great calorie confusion

There is an unopened packet of jam sandwich cream biscuits sat at my left hand. The nutritional information on the side of the packet tells me that the energy contained within a single delicious biscuit is 75 kcal, which I interpret to mean 75 kilocalories, or 75,000 calories. However, it also says that the same biscuit contains 312 kj, or kilojoules, and even more confusingly just above this it tells me each biscuit has just 75 calories. What on earth does it all mean?

In the end, all three – kilocalories, kilojoules and calories – measure the same thing, which is the energy contained within a single biscuit, but each is expressed with a different unit. The official scientific unit for energy is the joule, as defined by the International System of Units (see page 55). It was named after the wonderfully called and all-round clever chap, James Prescott Joule. The trouble is that energy comes in many different forms, each having its own special unit of measure. So, the unit of electrical energy is the kilowatt-hour and the unit of energy contained in gas is the therm. Horsepower-hours and British thermal units are used for cars and heating systems respectively, but my favourite is the delightfully named erg. The erg was the energy unit that was part of the now defunct centimetre-gram-second system, invented for no apparent good reason in 1873 by the British Association for the Advancement of Science. Sadly, when the centimetre-gram-second system was replaced with the eminently sensible metre-kilogram-second system, the erg was replaced by the less amusingly named joule.

For food, we currently use two systems: the calorie and the joule. The energy contained within food was initially gauged by thoroughly burning it within an enclosed container and measuring the temperature rise this caused in a small volume of water. This is where the calorie comes from, and 1 calorie was defined in 1824 as a unit of heat. It is the heat energy needed to raise the temperature of 1 g (about ¼ oz) of water by exactly 1 °C. So, 1 kilocalorie could heat up 1,000 g (2¼ lb), or 1 litre (1¾ pt) of water by 1 °C.

This is not, however, how most food energy measurements are made these days. While laboratories used literally to burn food to measure the energy it contained, they now use something called the Atwater system. With this method you first work out the total separate amounts of protein, fat and carbohydrate in the food you're testing. The energy contained in the food is then calculated using average values for the energy in protein, fat and carbohydrate. So, my biscuit contains 10 g of carbohydrate at 4 kilocalories per gram, which mean that 40 kilocalories in my biscuit are from carbohydrate. On top of this the biscuit has 1 g of protein at 4 kilocalories per gram, and 3.4 g of fat at 9 kilocalories per gram. Multiply this all together and add it up to give 75 kilocalories per biscuit.

Unfortunately, the calorie represents a rather small unit of energy. Most portions of food that we eat contain many thousands of calories: in my biscuit alone there are 75,000 calories, or 75 kilocalories. This is where the real confusion arises. Since the energy in all but the tiniest food portions is measured in kilocalories, it has become normal within the food industry to replace the term kilocalories with just calories. To make matters worse, this is done without any consistency. Which is why my biscuit is listed on the packet as containing both 75 calories and 75 kilocalories. It's even more confusing in European countries, where these energy measurements are also expressed in the more scientific units of joules.

It would be simpler if we could all just stick to the one unit and get rid of the bewildering mix of calories, kilocalories and joules. I'm not sure which I would prefer. The scientist in me wants to use joules, because those are the official International System of Units measure for energy. But the calorie is a much more down-to-earth unit. The definition that 1 calorie equals a raise of 1 °C in 1 g of water is something we intuitively understand. It gives us a way to get a handle on the energy content of the food we put into our mouths, which has become a critical issue with the Western diet. At the very least, we should get rid of the use of calories when we really mean kilocalories.

However, it occurs to me that I am really avoiding the issue that worries me most. As I write, I am deeply unnerved to realize that the two biscuits from the opened packet next to me could heat 2 litres (3½ pt) of water from room temperature to almost boiling point. I think it would be wisest if I put the packet away before any more biscuits go the way the first two did.

The dribblesome teapot

For the British population among whom I count myself, the taking of tea is part of our cultural heritage. You would think, then, that a dribbling teapot would be of particular concern to us. However, the definitive paper on the subject comes from a quartet of Frenchmen based at the University of Lyon. They discovered that certain features of the dribblesome teapot could not be predicted by the science of hydrodynamics (although it should be pointed out that this was because no one had ever bothered to look before).

According to the world pre-teapot experimentation, when large amounts of liquid (such as a mug of tea) flow through a tube (such as a teapot spout), the nature of the tube's surface should not matter. But we know it does. The traditional cure for a dribbly teapot is to smear a little butter on the lip of the spout, which always struck me a daft idea as I would rather lose half my tea to a dribble than drink butter-slicked tea. Clearly, the surface of the lip of the teapot is crucial to its dribble or non-dribble state.

The French team discovered that there are three things that come into play. First is the speed of the liquid flow. Faster liquid is less prone to dribbling, which explains why being careful and slowly pouring the tea just makes the dribble worse. Second is the radius of curvature of the lip of the teapot. A teapot with a thin, sharp-edged lip is much less likely to dribble than is a thick, gently curving, earthenware teapot. Which explains why metal teapots are usually better pourers. Finally, and this explains why the butter works, a water-repellent spout material guarantees to kill the dribble.

All three effects combine to create a dribble as follows: as a stream of tea shoots off the edge of the spout a little bit of liquid sticks to the very edge of the lip. If the lip is not very water-repelling, the tea sticks better to the lip and pulls the stream of liquid back and down towards the underside of the lip. The amount by which the stream is pulled back also depends on the contact angle, and this is determined by the curvature and thickness of the lip. Together, these effects can pull the stream of tea far enough back that it clings to the underside of the spout and dribbles. As the flow of liquid speeds up, the effect becomes too small to deflect the faster moving tea and the dribble stops or, at least, diminishes.

However, understanding the hydrodynamic causes behind teapot dribble is not much recompense if your own teapot is dribbly. You could try pouring the tea faster, but from my own experience this just ends up with more tea on the table as you overshoot the cup. Another trick, that I have seen in Chinese restaurants, is to change the thickness of the lip of your teapot. By pushing a short length of clear plastic piping over the spout and cutting it at an angle, you can make a new, thin, sharp-edged and more water-repellent spout. Except, of course, that it looks awful and ruins your fancy teapot's designer lines.

If neither solution appeals, all you can do is to apply a layer of super-hydrophobic (very water-repellent) material to the lip. The classic smear of butter does just this, but also leaves a thin oil slick on the tea. Modern science has a variety of super-hydrophobic materials that are unfortunately expensive and tricky to get hold of, except for one: soot. A layer of soot from, say, a candle, creates a surface that will not wet – any water just beads off. Hold your teapot spout over a candle flame until it is blackened, wipe off any soot from the top of the spout but make sure to leave soot on the inside of the bottom lip, and your teapot will be dribble-free. True, you will now get flecks of soot in your tea, but that has got to be better than butter.

Kitchen scales and the kilogram

Digital kitchen scales are, in my opinion, one of the twenty-first century's great gifts to humanity – they take up little space, are incredibly easy to use, can flick between imperial and metric units and you can zero them with any size bowl sat on top – and yet, they almost certainly lie.

If I plonk a lump of cheese on my scales and it says 153 g (about 5 oz), can I be certain that it really weighs 153 grams? If you look closely at the scales, or perhaps in the manual, there will be a degree of accuracy, which on my scales ranges from plus or minus 5 g (about ¼ oz). So, my lump of cheese could in reality weigh anything from 148 to 158 g (5¼ to 5¾ oz). This will probably not make much difference to my cooking, but it does raise the question of whether I can trust that even this range of weights is correct? Is it possible to ever know the weight of anything precisely and with 100 per cent certainty? The answer is yes, but only with one small object in the universe.

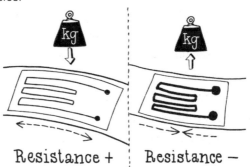

Resistance +   Resistance −

A foil strain gauge stretching and contracting

My scales were almost certainly made somewhere in the Far East. Inside the scales is a device called a strain gauge, which converts the weight loaded onto the scales into an electrical signal. Strain gauges consist of lots of parallel, incredibly thin, strips of metal foil. When the gauge is squashed by a weight, the foil strips are stretched even thinner. As they stretch, their resistance to electricity changes, and it is this that the microprocessor within the scales detects and converts into numbers on the display. During its construction in the factory, the microprocessor is calibrated so that it recognizes the strain gauge's reading for 0 g and 1 kg. From this, the microprocessor can work out the weight of anything placed on top of it. During calibration, the factory uses a test weight that is exactly a kilogram – but, of course, they only know it is a kilogram because they weighed it on a more accurate set of scales made in a different factory, and so it goes on that each weighing device is calibrated using a standard kilogram that was in turn weighed on a more accurate device. Since each set of scales works with an inherent degree of inaccuracy, each subsequent standard kilogram will produce an ever-larger variation in measured weight. So, where does it end? If you keep going back up this chain what do you find? Eventually, the calibration of my scales goes back through the Far Eastern manufacturer to a suburb of Paris, France.

In 1960, at the eleventh *Conférence Générale des Poids et Mesures* (General Conference on Weights and Measures), the gathered dignitaries announced *Le Système International d'Unités*, or SI Units, as it became known. This standard defined seven fundamental units and how to measure them. The system has been updated since then, and for all but one of the units the definition can be counted, with some difficulty, in nature. For example, the metre is now defined as the distance travelled by light, in a vacuum, in one 299,792,458th of a second. The second is the time taken for 9,192,631,770 cycles of the radiation coming from a specific type of caesium atom. The single irregular unit of measurement is the kilogram. The kilogram is defined, in absolute terms, as the weight of a lump of 90 per cent platinum and 10 per cent iridium made in 1889 and currently located in a vault in Sèvre, on the outskirts of Paris. Calling this a lump of platinum and iridium is somewhat underselling it, though: the International Prototype Kilogram, as it is called, is a perfect, flawless cylinder 39.17 mm tall (just over 1½ in), with an identical diameter of 39.17 mm. Copies of the IPK were made and distributed around the world, where national institutes of weights and measures use these first generation copies to make more, inevitably slightly less accurate, second generation copies – and so it goes on, all the way back to my kitchen scales. With each step away from the International Prototype Kilogram, the weighing device becomes more and more inaccurate. When my kitchen scales confidently tell me that a lump of cheese weighs 153 g, the chance that this is actually true is incredibly slim.

Cooking with science

One of the great pleasures of my life is cooking, and it probably won't come as much of a surprise if I tell you that I am also a fan of kitchen gadgets. My kitchen cupboards and drawers are crammed with all manner of utensils and devices, some useful and some less so. However, the biggest and most spectacular is my induction hob. This feels to me like a piece of science fiction imported from the world of *Star Trek*, or other space opera of your choice. The cooking surface is a completely smooth, black ceramic sheet, free of any visible means of heat production. Yet if I pop a saucepan of water on the hob and turn the dial, I create instant heat and can boil the water in mere moments. What makes it really incredible, though, is when you lift off the pan full of boiling water and put your hand onto the hob. So long as the pan has not been on the hob for very long, rather than a shriek of pain as your hand welds to the ceramic, the hob feels warm, but not hot. How can it possibly have boiled a pan of water without itself getting hot? What trickery is this that makes it work? And please, if you are going to try this, do take care.

Induction, or to give it its full name, electromagnetic induction, was first discovered, as so many electrical things were, by the great Michael Faraday, working in the basement laboratories at the Royal Institution in London. Wonderfully, we know not only the exact location of his discovery, but that it happened on 29 August 1831. As he put it in a letter to a friend, 'I have got hold of a good thing but can't say; it may be a weed instead of a fish that after all my labour I may at last pull up.' It turned out to be a whopper of a fish.

What Faraday had metaphorically caught was this: if you wiggle a magnet near a wire, an electric current will flow in the wire and, crucially, the reverse is also true. If you make a current flow back and forth in a wire, it produces a magnetic field around the wire. And that's it; that's pretty much all there is to electromagnetic induction. However, the consequences are far reaching.

Take a flat coil of wire and put a rapidly alternating current into this wire. According to electromagnetic induction, this changing flow of current produces a magnetic field that will itself be alternating with a North Pole, first one side then the other. If you put a lump of a magnetic metal such as steel on top of the coil, the alternating magnetic field will induce a current to flow in the lump of metal. Now, it turns out that steel is a bit rubbish at conducting electricity, as the flow of electricity is resisted, and some of its energy turns into heat. So our lump of steel on the coil of wire begins to heat up, even though the coil itself does not.

Now build the coil of wire under a stylish black ceramic sheet, and put the lump of steel in the bottom of a saucepan. Hey presto, you have made yourself an induction hob. The magnetic fields being produced by the coil of wire will work through the ceramic sheet. You can even try the trick used by the salesmen of the first induction hob developed by the Westinghouse Electric Company in 1973. They would cook food, to the wonder of onlookers, through several sheets of newspaper lying on top of the hob.

However, the nine-year-old boy sitting on my metaphorical shoulder has an annoying question that I have as yet not answered. Why does all this electromagnetic induction happen in the first place? What is it about electricity and magnetism that makes them go hand in hand? A hint of the answer lies in the word electromagnetic itself. It is not so much that the two are always linked together – rather, they are the same thing.

There are only four fundamental forces in nature. The weak nuclear and strong nuclear forces are what glue atoms together, the gravitational force we don't really understand but it gives us gravity, and then there is the electromagnetic force. This one force can be envisaged, in a completely arbitrary way as two forces at right angles to each other. One of these we call magnetic and the other electrical, but it is only our flawed perception that sees them as different. The reason electricity and magnetism go hand in hand is because they are fundamentally the same thing.

Which brings me back to my original point, that the induction hob is the most remarkable of kitchen gadgets. While I'm happy to be proven wrong, I'm fairly sure that no amount of slicers, popcorn makers, whisks, rice cookers or bread machines can so elegantly demonstrate one of the fundamental forces of nature.

The microwave inside-out myth

If you don't have one yourself, you certainly know somebody who does. Microwave ovens sit on nearly everyone's kitchen counter, cuboids of accelerated cooking, but what's going on inside and how do they really work? Search the internet and textbooks and you will be told that microwave ovens heat from the inside out. Furthermore, we are told that microwaves cook by causing water molecules to resonate. This is almost, but sadly not quite, true.

Back in 1945, a chap called Percy Spencer was working on a military project for the USA that involved microwave transmitters. It was the days before health and safety and, as Percy casually stood next to an unshielded transmitter, a chocolate bar (a Hershey's Mr Goodbar) that he had in his pocket melted. Little did Percy realize at the time, but this was the first instance of microwave cooking. Percy was lucky he didn't cook himself.

Microwave rays are part of the electromagnetic spectrum, which is to say that they are the same as light rays. What makes them different is the distance between peaks of the wave. In the case of the microwaves from an oven the distance between peaks is 12.2 cm (4¾ in), while visible light is about 200,000 times smaller. A crucial part of the electromagnetic microwave is the electro bit. Part of the wave is an electrical field that goes up and down with the wave, from positive to negative.

Now, imagine a molecule surrounded by microwaves. If this molecule has one side that's a bit more positively charged than the other, it will try to align itself to the electrical field

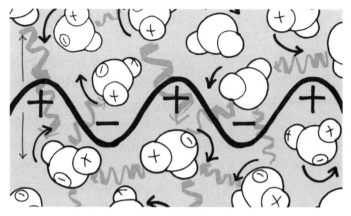

Molecules flip-flop with the microwaves

of the microwave. Since this field goes up and down, the molecule tries to flip-flop with the wave. As a consequence, some of the energy from the microwave is transferred to the molecule. What's more, as the molecule flip-flops, it bashes into other molecules around it and generously passes on some of its newly found energy. This type of energy transfer is called dielectric heating and, as the name suggests, it's just a way to make things hot.

If our theoretical molecule is a water molecule in a theoretical bowl of theoretical vegetable soup, it has an uneven spread of electrical charge, so will happily do the microwave flip-flop dance. Before long, the water molecules in our soup are jostling about, grabbing energy from the microwaves, passing it around and heating up the soup. However, it's not just water that jiggles; both fat and sugar can do this, too. Anything with an uneven electrical charge will heat up – even pottery plates will warm in a microwave if the glaze contains something with an uneven charge. What's more, this process has nothing to do with anything resonating, let alone water. Interestingly, water molecules in ice don't do the flip-flop dance as they are not so free to move, which is why thawing frozen food seems to take forever in a microwave oven.

So, if that's what makes the microwave oven generate heat, how does it heat from the inside out? Well, it doesn't – this, too, is a fallacy. Microwave ovens cook from the outside in, just like all ovens, but the microwaves do penetrate into your soup, or potato, or leftover curry, by a couple of centimetres (¾ inch). It is this that speeds up the cooking time. Rather than just heating the surface and the heat slowly being passed inwards, the heat is injected a little way into the food, giving it a head start. You also don't need to warm up the microwave oven in advance, like you do with a regular oven. The microwaves start heating instantly, and you don't lose any energy to the walls of your oven. Metal plates act like a mirror to microwave rays: they bounce them off and back into whatever you are trying to heat. Taken all together, that's why your microwave oven cooks so quickly.

It has been over sixty years since Percy Spencer's bar of chocolate melted in his pocket. Since then microwaves have been used to cook countless items all over the world, yet the microwave oven itself remains a somewhat mysterious and misunderstood contraption.

The imperfect toaster

Despite the promises of numerous manufacturers of small kitchen appliances, I have yet to own a toaster that really works. Even without touching the settings, my toast can come out with a huge range of toastiness, from faintly tanned to thoroughly carbonized. Maybe I just keep buying cheap toasters, or maybe there is something fundamentally difficult about automating the toasting of bread.

The basic design of the toaster has essentially remained unchanged since 1919, when Charles Strite patented the automatic, pop-up toaster. Strite's invention brought together a number of ideas in one machine: notably, a heating element on a timer, linked to a spring-powered pop-up mechanism. At the heart of the toaster, though, is another invention, that you can still see glowing brightly in a modern toaster: nichrome wire. The very first toaster, invented in 1893 by Scotsman Alan MacMasters, used coils of steel wire, through which electricity flowed, to produce heat to toast the bread. Unfortunately, the steel wire would overheat, react with oxygen and burn away. The company manufacturing the toasters, and MacMasters himself, did not do well from his invention.

Then, in 1905, nichrome wire came onto the scene. If you make a mixture of 80 per cent nickel and 20 per cent chromium, the resultant alloy has a couple of very important properties. First, it can be heated to very high temperatures without burning away like steel; instead nichrome forms a protective layer of chromium oxide. Second, nichrome turns out to be a slightly rubbish conductor of electricity. You may think that this is a hindrance for use in electrical devices. However, it is just this resistance to the flow of electricity that makes nichrome wire essential to most electrical heating appliances. When electricity runs through nichrome wire, its resistance manifests as heat, and lots of it. These two properties combined make nichrome the ideal material to convert electricity into heat; so much so that its American inventor, Albert Marsh, was magnificently declared the father of the electrical heating industry.

So, if the toaster itself is such an essentially simple device, why does it continue to produce such variable results? The answer lies not in the toaster, but in the bread. The perfect toast, in my estimation at least, is hot, crispy and golden brown all over. The hot and crispy bit is relatively straightforward to achieve, but the colouring is more tricky. The chemistry of this change, called the Maillard reaction, has been studied extensively, as it is fundamental to many cooking processes. As you heat up a slice of bread (or perhaps a potato, a coffee bean, or a steak), protein molecules begin to react with certain sugars, such as glucose, lactose and maltose – but not sucrose. This reaction produces new, complex, brown-coloured and very tasty molecules. It is these molecules that we strive to make on the surface of our toast. Heat it too much, though, and you take the reaction too far, producing bitter-tasting caramelization and, ultimately, carbonization.

The problem with making toast is that the extent of the Maillard reaction critically depends on the amount and type of sugar and protein in the bread. This is why even the best currently available toaster in the world will be unable to reliably produce a perfect slice of toast every time. Even between similar loaves of bread there will be enough variation to make a difference. Furthermore, purely physical aspects such as the temperature of the bread before it goes in the toaster and the thickness of the slice will have a major impact on the Maillard reaction. It turns out that making toast is harder than it seems, which might be why the evolution of toaster technology has stagnated for nearly a hundred years.

The coffee-ring conundrum

If you spill some coffee on the countertop in your kitchen and leave it to dry, it does not, as you may expect, leave behind a uniform splat of brown. Instead, you end up with a very dark ring along the edge of where the coffee spilt, but inside this ring will be relatively free of the brown stain. You get the same effect, to a lesser extent, on napkins and tablecloths stained with red wine. As the stain dries it becomes more intense along the edge of the liquid.

This is known as the coffee-ring effect, and it's worth noting that this name comes from the edge effect when a drop of coffee dries out rather than the shape of a stain from the bottom of a dribbly mug of coffee. It happens because coffee is not just a brown liquid, but a suspension of tiny ground-up

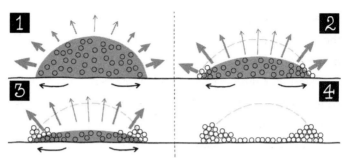

How spilt coffee evaporates to give a coffee ring

particles of coffee bean in water, along with dissolved flavour molecules – and if you have it caffeinated, the caffeine that I personally regard as essential.

Imagine we have a droplet of coffee sitting on a hard countertop in your kitchen. As our droplet begins to dry out, you'll notice a couple of things. The size of the area of wetted countertop does not get smaller as the droplet evaporates. The edge of the liquid on the counter stays locked in position, because water is very good at sticking to things: to such an extent that the forces pulling the droplet together are weaker than the ones holding it to the countertop. So, as the droplet dries, it doesn't get smaller in diameter; it gets flatter.

The second thing going on is that water evaporates across the whole surface area of the drop, including the edges where the water curves down to meet the countertop. As water molecules evaporate from the middle of our drop, they are replaced by water molecules from below. The situation at the edge is a bit different – here the water is angled down to the countertop. So, when molecules evaporate from the surface at the edge they are replaced by water drawn from the centre of the drop. This creates a current from the middle of the drop towards the edge.

Since our drop is packed with tiny ground-up specks of coffee bean, these are pulled along with the current. They begin to accumulate as a coffee ring and by the time the water has completely evaporated, the majority of the particles have piled up around the edge. In the case of a spill on an absorbent surface, such as a napkin, the same thing happens. You don't get such a clear coffee-ring effect only because the flow of particles is hindered by fibres of napkin.

This may seem like a very domestic problem, but it is a serious issue within the paint industry. The coffee-ring effect applies to any liquid containing a suspension of tiny particles. In a can of spray paint you have exactly this: tiny fragments of pigment suspended in a carrier liquid. What you want to achieve with paint is an even coating, and not little rings of dark colour caused by the coffee-ring effect. There are a few ways you can get around this problem. The simplest is to use a carrier liquid that evaporates so quickly that the particles don't have time to move within the liquid.

More intriguingly, scientists at the University of Pennsylvania, in the USA, found that if your particles suspended in the liquid are elongated, rather than spherical, you don't get a coffee ring. If the particles are about three times as long as they are wide, they get stuck on the inside surface of the droplet. They then start sticking to each other and make clumps that are too big to be drawn towards the droplet edge. When the droplet dries, you get a smooth and even distribution of particles and this may be the way forward for slow drying spray paints.

So, to avoid unsightly, dark coffee rings on your countertop you can either make sure to grind the coffee into elongated particles, or you can wipe up the spill before it dries. While one of these solutions may have the scientific high ground, the other is probably simpler.

Your ice cubes are not doing what they should

The chink of ice cubes in a tall glass of gin and tonic conjures memories of hot summer evenings for me, but if you're not a fan of G&T, please substitute a chilled beverage of your choice. Irrespective of the drink, something very peculiar takes place in your glass in which the ice floats.

Consider for a moment the basic difference between liquids and solids, and specifically I want to use pure alcohol, or ethanol as an example, simply because it's a very well-behaved substance. The molecules within liquid ethanol are held together loosely and are free to move around. This is the basic definition of a liquid, and it allows us to pour liquid ethanol to fill a container. If you freeze ethanol to −114 °C (−173 °F), it turns into a solid. Inside this solid ethanol the molecules have become locked into place in a neat and regular crystalline array.

Since they are no longer so free to move about, the molecules in solid ethanol pack tighter, each taking up less space, and the solid ethanol is denser than the liquid ethanol. If you made cubes of solid ethanol and dropped them into a glass of liquid ethanol they would sink to the bottom.

This is the case for virtually all substances, whether you are talking about well-behaved substances such as ethanol, cooking oil, mercury, oxygen or steel. The solid is denser, and sinks in the liquid. Water, on the other hand, is different and positively awkward: ice is less dense than water, and floats.

This peculiarity is all down to the ability of water molecules to form a special type of relatively weak bond called a hydrogen bond. Water is particularly good at making hydrogen bonds, and it's what gives water some of its other odd characteristics, such as surface tension and capillary action. In liquid water, the molecules are zooming around with too much energy for hydrogen bonds to lock them into place. Consequently, the molecules can move relatively close to each other.

When the temperature drops below 0 °C (32 °F), the molecules can't resist the hydrogen bonds, and they are slowed to a halt. The molecules arrange themselves in layers of hexagonal arrays, with the spacing determined by the length of the hydrogen bond. The combination of this specific geometrical arrangement, and the length of the hydrogen bond, pushes the water molecules away from each other when compared to liquid water. With fewer molecules of water packed into a given space, the density goes down rather than up.

While it may seem trivial whether your ice cubes float or sink, this peculiarity of water has a profound impact on our world. The vast Arctic ice cap, with all the polar bears and arctic foxes living on it, floats over the North Pole, rather than lying at the bottom of the ocean. It's not clear what the effect would be if this wasn't the case. If the Arctic ice sank, one scenario has a gradual accumulation of ice at the bottom of the sea, which in turn cools the water above and the whole atmosphere; more ice forms and so on until the oceans are entirely ice, the world turns into a giant snowball and we all die. While I will admit that this may be a smidgeon melodramatic, at the very least winter-time ice at the bottom of every lake and stream would kill off vast swathes of the beetles, bugs and crustacea that live down there.

Clearly, the density of frozen water is never going to change. It's a fundamental physical property, a product of the specific chemistry of water. Neither is this floating quirk of ice lucky happenstance. Rather, it's an elementary driving force of the evolution of life on our planet. If ice didn't float, we almost certainly wouldn't be here to consider the subject. All of which is probably more than you want to consider while listening to the chink of ice cubes in your drink. I suggest you put it from your mind to enjoy your beverage, before the freakily floating ice melts away.

The wonderful wax candle engine

Take a match, strike it and hold the flame to the wick of a candle. Within moments the candle will flicker and an orange flame will spring into life. Now leave the candle burning for a while and over time it will get shorter. Clearly, the wax is being consumed by the flame and used as a fuel. However, now take a second candle, a second lit match, and try to set fire to the body of the candle. It can't be done. No matter what you do, you will not be able to set fire to the wax that makes up the candle itself, yet you can light the wick. Incredibly, wax is non-flammable.

This seemingly paradoxical observation gave rise to one of the earliest and most elegant of popular science books, *The Chemical History of a Candle* by Michael Faraday. This book is made up of the collected notes from a series of six lectures given by Faraday in 1848 as part of the Royal Institution of Great Britain's annual Christmas lecture series that still take place to this day. Faraday himself was a brilliant scientist who discovered several chemical elements, invented the electric motor and, arguably, is the father of the popularization of science. Faraday had a mind that could take the simple yet scientific observation of a candle to amazing lengths, by knowing what questions to ask.

Wax at room temperature is an non-flammable solid. When you see a candle flame, what you are seeing is the combustion of the gaseous form of wax, or wax vapour. Maybe this doesn't come as a surprise, as the flame itself is clearly neither solid nor liquid and so it must all be happening in the form of gases. What makes the candle remarkable is that it is a beautifully elegant engine for converting solid wax to a gas and then combusting it.

The wick of a candle is usually made of a braided cotton material, which will not burn particularly well on its own. With a fresh candle, it will burn just well enough to produce heat that radiates to the solid wax beneath. This melts the wax at the tip of the candle, turning it from a solid to a liquid, and the liquid is then drawn up the candle-wick by capillary action. As the liquid wax gets closer to the heat of the burning cotton wick it vaporizes, turning from a liquid into a gas. This hot vapour begins to rise, drawn up by the convection of air around it, into the flame of the burning wick. Now we have wax vapour in the presence of plenty of oxygen from the air and a source of ignition from the burning wick. The wax vapour combusts, producing a much bigger flame, with much greater radiated heat, more of the candle's solid wax melts and is drawn up the wick which keeps the candle burning. The engine that is the candle has begun and will continue until the wax runs out or the flame is blown out. Put like this, it seems simple, but each step has wonderful subtleties.

Capillary action is a peculiar phenomenon where the tendency of the molecules in a liquid to stick to each other, combined with how well they stick to other objects, allows the liquid to pull itself up by its own boot straps. For capillary action to work you need to match the physical properties of the liquid, its surface tension and density, to the thing it's being drawn up. In the case of a candle, there are narrow gaps between the strands of cotton in the wick. These gaps are just the right width for the liquid wax to be drawn up, which is why wicks are pretty much always made of cotton. Other materials produce differently sized gaps that won't work so well. It's also why all wicks on partially burnt candles are about the same size. It's the non-flammable liquid wax in the wick that stops it being completely consumed and burnt away by the flame. Since the height that wax can be drawn up a wick is determined by capillary action, which is about 1 cm (about ½ in), that is how long your wick gets.

The shape of the top of the candle is also critical for successful burning. When a candle has been alight for a while, the familiar pool of wax develops at the top. This pool forms a reservoir of liquid wax, ready to be drawn up the wick to be vaporized and burnt. If you don't have this pool, either because you fiddled with the candle or because the candle is too narrow, not only have you wasted the wax but also the candle will not burn so well. It will have a smaller flame that is more prone to guttering and going out because there is less wax being drawn up the wick. The advantage of this pool of wax is why candles tend always to be made with a minimum diameter of about 1 cm. Smaller candles than this, such as those sold for birthday cakes, don't develop a pool – they just produce lots of molten wax that dribbles down the sides.

The flame of a candle is worth a closer look, too. Immediately above and around the wick is an area that is darker

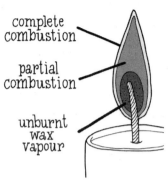

complete
combustion

partial
combustion

unburnt
wax
vapour

than the rest of the candle. This is unburnt wax vapour without enough oxygen mixed in with it to combust. As this wax vapour rises, more oxygen diffuses in and we get to the glowing yellow part of the flame where the wax begins to burn. But this area of the flame still lacks enough oxygen and so the wax is incompletely burnt, leaving some of the carbon from the wax as carbon particles rather than carbon dioxide gas. All that unburnt carbon gets very hot – yellow hot, in fact, which is why the top of a candle flame is the colour it is. There is a third part to the candle flame, though it is very tricky to see. Around the outside of the candle flame, sheathing the yellow part is an almost invisible blue-yellow layer about 2 mm ($1/16$ inch) deep. To see this layer, try setting a candle against a dark background and illuminate it from the side: look carefully along the vertical edges, and you should be able to detect a difference in the outside of the flame. This is the area where there is sufficient oxygen for complete combustion of the candle wax to take place. It is also the hottest part of the flame.

One simple demonstration, that dramatizes something of a candle's properties, never fails to delight. To begin, light a candle and burn it for a while, so that the flame is big and stable. Then, with a lit match in your hand, carefully blow out the flame. You will see a stream of what looks like smoke rising from the extinguished wick, but it's not smoke, it's wax vapour. Now quickly bring your lit match to within a few centimetres (about 1 in) of the wick and into the stream of wax vapour. As you do, a flame will jump from the match, along the wax vapour, and the candle will be lit once more. When you have mastered this, try using a candle-snuffer so that the flame can be extinguished with the minimum of air disturbance and the wax vapour settles into a vertical stream more quickly. With practice, you can get the candle to relight itself from up to 5 or 6 cm (a couple of inches) away.

All of this science, and much more, is detailed in Faraday's wonderful *The Chemical History of a Candle*. He filled the book with experiments that investigate the seemingly simple but, on closer examination, complex science of a candle. The book is still in print and digital copies are available for free; it's definitely worth a read.

The Marvels of Science Around the House

Lighting up slowly

It began as a trickle but has become a flood, as people around the world chuck out their old incandescent light bulbs in favour of new-fangled compact fluorescent ones. Governments around the world are passing legislation that prohibits the use of incandescent light bulbs. Brazil and Venezuela started the trend back in 2005, with Australia in 2010 and the United Kingdom in 2011 following suit. At the time of writing, Russia, the USA and China are all undergoing the same process. The reason for this is simple: the incandescent bulb is a horribly inefficient producer of light. It has only lasted as long as it has due to a lack of economically credible competitors.

The traditional incandescent light bulb was first demonstrated in practical use not, as is commonly believed, by Thomas Edison, or even by Joseph Swan, but by James Bowman Lindsay in 1835, in Dundee, Scotland. While the invention has been massively improved, nearly two hundred years later only about 2 per cent of the energy you put into an incandescent light bulb is turned into visible light. Compare this to the light bulb to which we are all switching, the compact fluorescent light bulb that converts about 10 per cent of its energy into light, and it's easy to see why we are all being encouraged to make the change.

The compact fluorescent light bulb is essentially just a fluorescent tube coiled up and in some cases enclosed within an outer bulb of glass. The science behind the basic operation has been known since 1856, but it was the innovation of coiling and miniaturization in 1976 that started their journey into our homes. The tube of the compact fluorescent light bulb is filled with inert argon gas at very low pressure, but there is also a tiny drop of liquid mercury inside the tube, which heats up and vaporizes when an electric current passes through the tube. As it does, the electricity passes some of its energy onto the mercury atoms. The mercury can only hold onto this energy for a very short time, before quickly releasing

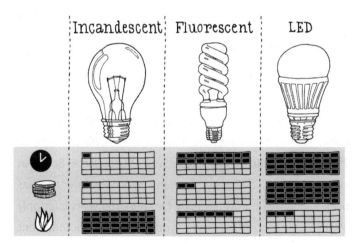

How modern light bulbs compare

it in the form of invisible ultraviolet light. The ultraviolet light then hits the white, powdery phosphor coating on the inside of the glass tube, which absorbs energy from the ultraviolet light and, just like the mercury, quickly gives up the energy, but this time as visible light. The light produced from modern compact fluorescent light bulbs is pretty much the same colour as light from an incandescent bulb. Where the technology is still catching up is in the comparatively long time that it takes for fluorescent light to come to full brightness.

It normally takes a compact fluorescent light bulb between ten seconds and a minute to reach its full brightness, and here's why. Before you turn on a fluorescent bulb, inside the tube there is very little mercury gas; it's almost all in a liquid state. What's more, the argon gas inside the tube is not conductive to electricity. To get the electricity flowing along the tube you need a tiny coil of wire at each end. As electricity flows through these wires, they heat up and begin to fire electrons off their surface into the argon gas. This then heats up the mercury, turning it into a vapour, and only when this reaches a critical point does electricity begin to flow in earnest through the tube. The mercury can then give out ultraviolet light that the phosphor converts to visible light. All of this takes a little while to get going, which is why ordinary compact fluorescent light bulbs don't work so well outdoors. If it's cold it can take up to five minutes for them to become fully bright.

There have been some recent innovations that may lead to quicker start-up times for compact fluorescent bulbs, but the technology will never have the instant illumination provided by incandescent bulbs. Despite this, the fivefold increase in efficiency and resulting enormous energy savings more than make up for the inconvenience. However, waiting in the wings is yet another technology, the light emitting diode. Currently, light bulbs made from these are significantly more expensive, but the efficiency is double again that of compact fluorescents and they give you instant light no matter what the temperature. While the compact fluorescent bulb is pushing out the 200-year-old incandescent, it may not keep its place in the spotlight.

End over end down the stairs

In 2014 I had the chance to attempt to set the Guinness World Record for the most stairs descended by a Slinky. With the help of Hugh Hunt, an engineer from Cambridge University in the United Kingdom, we successfully set the record at thirty steps. In the process – and I should mention that it is harder than you may imagine – I had pause to wonder how a Slinky worked in the first place.

The Slinky was invented in 1943 by Richard James, an engineer based in Philadelphia in the USA. The original design, that endures to this day, is a coil of steel over 21 m (69 ft) in length, with 98 loops of wire. When it went on sale in 1945 it was an immediate success, apparently selling out the initial production run in just ninety minutes. Since then, hundreds of millions of Slinkys have been sold, and that's not including the modern plastic versions.

The magic of the Slinky comes when you lift and turn the top of the Slinky over the edge of a step, the whole coil spools down onto the step lower down, then once on the lower step, the Slinky automatically starts again and descends to the next lower step, and so on, until it gets to the bottom or, more often, gets itself tangled and stops. It looks like it shouldn't work, but it clearly does.

Every spring, no matter the size, has what is known as a spring constant, which is a measure of not just how strong but also how long it is. It's critical to get the spring constant just right in a Slinky, and to match it to the steps you are attempting to descend. If the spring constant is too high, the Slinky will flip itself down the steps and start falling rather than slinking. Too low, and when the top of the Slinky reaches the next step down, it gets stuck and doesn't have enough pull to bring the bottom of the Slinky down as well. Similarly, if the size of the steps is wrong the Slinky won't work. With very shallow steps, for example, most Slinkys will get stuck, as there is not enough room to pull the whole coil down to the next step. Matching your Slinky to your steps is your first job.

The spring constant explains why the Slinky gets down the first step, but not why it keeps going. To get to the bottom of this, you need to watch the Slinky very carefully, ideally in slow motion, as it descends a staircase. What you will see is that when a Slinky gets to a lower step, the last few coils of the Slinky do not join up with the rest of the coils and do not, even momentarily, come to a complete stop. The momentum of those last coils is able to overcome the force pulling them down onto the Slinky. Consequently, they flip over the top and start to fall down the next step. Gravity takes over and the whole process begins again.

So, with the physics working in your favour, once you have the right steps for your Slinky, it should be possible to get it to walk all the way down. Even so, from my own record-breaking experience the secret to a really long descent is all down to the initial flick that you use to set the Slinky on its way. Get this right and your Slinky will keep going until it runs out of steps.

Machines that can see in the dark

In the corner of the room in which I sit as I write this, up near the ceiling is a small white plastic box. It's part of my burglar-alarm system and the front of this box has a curved opaque white plastic sheet. The box is oblivious to my presence, but if I get up from my seat a little red light comes on. Somehow the white plastic box sees me, even though I am at least 5 m (16 ft) away. Then, if I stand absolutely still, after about 5 seconds the light goes out. While it is possible to move slowly enough for the red light to remain off, it's incredibly difficult and the detector is amazingly sensitive; the slightest

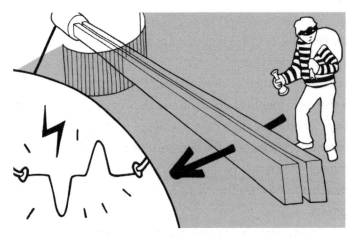

Gallium nitride in a PIR detector seeing an
intruding infrared source

rapid movement and the detector notices me. What's more, it
detects movement both in full daylight and in pitch-black. How
is it that something so small and innocuous can perform the
remarkable feat of recognizing me against the background?

The plastic box with the winking red light is known in
the burglar alarm trade as a passive infrared detector or PIR
detector. As the name suggests, these can detect infrared light,
which is essentially a colour of light that we can't see. Our eyes
can only detect the particular wavelengths of light that make
up the rainbow of colours. However, there is a continuous
spectrum of light with wavelengths that extend beyond these
colours in both directions. Light with wavelengths just shorter
than the colour red make up what is known as infrared light.
While we can't see this, in some cases you can feel strong
infrared light as radiated heat.

All objects emit infrared light as faint radiated heat and the PIR detector uses a thin crystal of a substance called gallium nitride to detect this. Crystals of gallium nitride have the unusual property that when infrared light hits them, it causes a change in the structure of the crystal. Accompanying this is a tiny change in the electrical properties of the crystal, allowing a minuscule difference in the amount of electricity that can flow through the crystal. While this is a very small effect, it can be detected using straightforward and readily available electrical circuits.

To see something moving within a room you need not one, but two tiny rod shaped gallium nitride crystals. These are arranged vertically in the detector, and placed next to each other with a sliver of space in between. Each of these crystals effectively sees out into the room along a thin vertical strip. Since these strips are really close together, the stationary background infrared light hitting each crystal, and the voltage produced by each crystal, is almost the same.

The really cunning bit is that you wire up the two crystals back to back so that the positive end of one crystal is connected to the positive end of the other. If the voltage produced by each crystal is the same it cancels itself out and overall you get no voltage. This trick effectively blinds the detector to things such as radiators, heat vents and other sources of slow changing, background infrared light.

If you walk across the room, you pass through the two thin strips the crystals are seeing. At some point you will momentarily be more in one strip than the other. When this happens the infrared hitting one crystal goes up, the voltages produced are no longer equal and they no longer cancel each other out. Suddenly, you have a spike in voltage from the pair of crystals that the detector can notice. What's more, since larger things moving across the strip create larger spikes in voltage, you can set the detector to ignore smaller pet-sized objects.

The problem with this system as it stands, is that it can only sense people moving in two narrow strips stretching out from the detector. To give the PIR detector a better field of view an array of plastic lenses are arranged in a curve around the crystals. Since we are only interested in infrared light, the plastic does not have to be see-through to visible light, only the infrared. So, even though the front of a PIR detector looks white and opaque, to infrared it's transparent. These plastic lenses focus strips of infrared light from a range of different angles onto the crystals. The detector as a whole can now see into the room in a half-dozen or so different directions at the same time.

All of which adds up to a nifty passive infrared detection system that ignores not only small creatures but also really slowly changing background infrared sources. While PIR detectors are blind to these things, they are exquisitely sensitive to unwelcome human visitors to your home, or in my case, people trying to see how slowly you need to move to outwit the burglar alarm.

Making glass one-way

Have you ever sat by a window watching the world go by while gradually it turns dark? At some point you find that instead of watching the passers-by, you are looking back at yourself. The previously transparent window becomes reflective as the world outside darkens. Clearly the glass has not physically changed, but it has turned into a mirror, at least from your perspective. If you go around to the other side of the glass, outside into the dark, and look into the illuminated room, the glass is once again transparent.

The key to what is happening here is to realize that glass is not as transparent as we assume. If you shine a beam of light straight at a pane of clear glass, about 4 per cent of the light is reflected directly back from the front face of the pane. What's more, the light is also reflected back from the inside surface of the other side of the pane of glass. In total, nearly 7 per cent of the light shining on the glass is directly reflected. Glass always acts as a mirror, just not a very good one.

Reflection can happen whenever light tries to pass from one medium to another. In the case of a window, the light is going from air to glass. A beam of light is essentially a wave of electromagnetic energy, which is to say that part of the energy is electrical and part magnetic. The surface of the glass is packed with electrons that are free to move about a little. The electrical wave part of the light causes these electrons to jiggle about, which in turn makes a magnetic field that jiggles about. All this jiggling magnetism and electricity manifests as light emitted

85

by the glass itself. Crucially, and as a consequence of the way they are made, the waves of this emitted light are exactly out of sync with the incoming light beam. Some of this light travels in the same direction as the original beam of light, but rather than adding to the beam, it cancels out a little bit of it. At the same time, and with the same intensity, the glass emits light going back where the original beam came from. The upshot of all this is that a small amount of our beam of light seems to bounce off the surface of the glass, while the rest carries on, somewhat diminished but unaffected. Ultimately, this is at the heart of all reflections, and is why glass acts as a mirror.

It does not, however, explain why you can't see this reflection during the day, but you can at night. For this you need to turn to biology. Our eyes are incredibly good at coping with different lighting conditions. They can adjust in a fraction of a second and we don't even notice it happening. Primarily we do this by automatically changing the size of our pupils that allow light into our eyeballs. By contracting or relaxing the muscles connected to our irises, our pupils shrink or grow. If your pupils are wide open, more light floods into your eyes and you can see in lower light levels; conversely, shrink the pupils down and you can see in bright conditions without your eyes being over-exposed. There are further mechanisms at the backs of your eyes, in the retinas, that gradually change the sensitivity of the light-detecting cells, but this can take thirty minutes.

During the day, sunlight streams in through your window. Even when it's cloudy your pupils are closed down, allowing only a small amount of light into your eyes. The light that reflects off the window from where you stand inside is comparatively very faint. Since your vision is adjusted to cope with high levels of light, you just don't perceive this faint reflection. It's there, but your eyes can't pick it up.

Conversely, at night if you are staring at a window, there is not much light coming to your eyes through the window. Your pupils will be wide open and capable of detecting the faint reflection. Go outside and look in through the window, into the bright room you were just standing in, and your pupils get smaller, your eyes adjust to see bright things, and the reflection disappears again.

Of course, if you turn off the light in a room, so that it's dark both inside and outside the room, there is now no light on either side of the window and you can't see anything.

Disappearing down the plughole left and right

If you travel to a country such as Ecuador or Kenya, both of which lie on the equator, you may be able to glimpse a classic performance of the Coriolis effect. Unfortunately, I have not had the chance to experience this in person, but have vicariously lived through it while watching Michael Palin's *Pole to Pole* television series back in 1992. When Palin reached the outskirts of Nairobi in Kenya, an enthusiastic young man showed him how water swirls anti-clockwise down a plughole in a domestic sink just north of the equator and clockwise just south of the equator. It's a common enough idea, and the explanation given for this observation is that it is caused by the Coriolis effect, that is in turn caused by the rotation of the Earth. The science behind this explanation is completely spot on and it is possible to do this demonstration. Unfortunately, there are more prosaic explanations for what happens in a domestic sink.

Named after a French mathematician, the Coriolis effect is a real thing that is most often encountered in the field of meteorology. It happens when something, such as air for example, moves across the surface of an object that is rotating, such as the Earth.

Imagine that there is a person sat in a stationary spaceship above the Earth and they have some newfangled bit of kit that allows them to monitor air currents. A current of air that is, from their perspective, moving in a straight line is in fact moving along a curved path on the surface of the Earth. The rotation of the Earth, and the friction between the air and the Earth, pushes the air to one side and makes the air current bend. In the northern hemisphere, which rotates anti-clockwise, air moving across the Earth's surface is pushed a little to the right. This means that as air blows inwards to an area of low pressure it begins to spiral around and to the right, in what becomes an anti-clockwise direction.

If you now jump to the southern hemisphere, this appears to rotate clockwise when viewed by our astronaut. The Coriolis effect is consequently reversed, and south of the equator air spirals into low-pressure areas in a clockwise direction. These spiralling cyclones of air create large-scale movements of air that drive most of the weather on the Earth. The most obvious of these large movements are hurricanes, rotating anti-clockwise in the north, and clockwise in the south. The Coriolis effect is a significant force where the movement of air involved is over large distances when compared to the size of the Earth, and in a time scale longer than the rotation of the Earth.

On smaller scales, for example in a sink, observing the Coriolis effect is a little tricky, but not impossible. In 1962, a Professor of Engineering at MIT in Cambridge, Massachusetts, built a huge, perfectly circular sink nearly 2 m (6 ft) across and 15 cm (6 in) deep. It was filled with water and left to settle for 24 hours. The sink was covered to prevent draughts stirring up the water, and the room containing it was carefully controlled at a uniform temperature. When the plug was pulled, it took twenty minutes to drain the sink, but when it did, and they tested it multiple times, it always drained anti-clockwise, precisely as predicted by the Coriolis effect. Clearly, you need to go to quite extraordinary lengths to see the Coriolis effect at smaller scales.

Given this, why then does my sink always drain clockwise, even though I am in the northern hemisphere? The answer is down to the shape of the sink, the water pressure in the cold tap and the fact that in almost all countries it is standard to fit the cold tap on the right hand side of the sink. Since the pressure in your cold water system is very probably greater than the pressure in the hot, when you fill a sink with both taps,

89

the cold water puts a spin on the water from the right in a clockwise fashion. When you pull the plug, this spin remains and a vortex forms also going clockwise. There will be a Coriolis effect opposing this clockwise vortex, but it's tiny.

It turns out that unless you have a vast sink with absolutely no pre-existing water currents, the Coriolis effect is so small that it cannot be detected in the average sink or bathtub. The distance that the water moves from the rim of a sink to the centre, is vanishingly tiny compared to the size of the Earth. Furthermore, the movement takes place over the course of a minute or so, which is also a minuscule fraction of how long it takes the Earth to turn.

In which case, what did Michael Palin and other tourists see demonstrated at the equator? In this case, it would appear that the bowls of water have been, how to put it, encouraged to swirl in the predicted direction. The trick, and you can try this in your own sink, is to carefully pour the water into the basin slightly to one side. This ensures the water has a gentle yet imperceptible spin on it. By changing from which side the water is poured, the direction of swirl can also be altered. Pull the plug and hey presto, you can make a basin empty clockwise or anti-clockwise at your whim. So, while the Coriolis effect is a real thing that is responsible for much of our weather, the only way to recreate it in a hand basin is with a little sleight of hand.

Einstein, relativity and your phone

Einstein is rightly famous for a number of extraordinary things. There is, of course, the great hair that he sported as he grew older, but probably more important than that is his work on relativity. Taken together, special relativity and general relativity become a unified theory of relativity that explains how time, gravity and velocity intersect. Except that you only get to see the effects of relativity across vast distances, or when travelling close to the speed of light. At least, that is our assumption, but in your pocket, or wherever your mobile phone happens to be, is a device that neatly demonstrates both parts of Einstein's genius and his theory of relativity.

Inside almost all smartphones is a tiny chip attached to an integrated antenna that calculates exactly where the phone is on the surface of the world to within about 3 or 4 m (about 10 to 13 ft). The global positioning system, or GPS, that allows it to do this relies not only on a network of satellites orbiting the Earth, but also on an understanding of relativity.

The job of a GPS satellite is, on the surface, pretty simple. Every thirty seconds it transmits a radio signal that contains not only the time the message was sent, but also information about the satellite's exact position above the Earth. The first of these, the time of the transmission, is taken from an atomic clock on board the satellite that is unfeasibly accurate to within 1 second every 138 million years. Knowing where the satellite itself is located is also not particularly difficult. Since it is orbiting the Earth above the atmosphere, its movement is very predictable using basic laws of motion. Even so, all GPS satellites are constantly monitored using ground-based radar to provide any teeny, tiny corrections needed. All this information is packaged up and transmitted to the world below every minute and half minute, precisely.

When your phone receives this signal, on its own it's useless. It needs to collect signals from three different satellites, over the course of thirty seconds or so, before it can start on a complicated bit of maths called trilateration. Note that this process is not the same as triangulation, as your phone doesn't know the angle the satellite signals are flying in from. What your phone does know though is at what time the signal arrived at the phone, since the phone has its own clock, too. By looking at the difference between the time sent and the time received, you can work out how long the message took to reach the phone.

Since we also know that the radio message travelled at the speed of light, your phone can work out how far away is the satellite that sent the signal. Once it has done this with three satellites, and knowing exactly where those satellites were when they sent the message, it's possible to work out the location of the phone using the maths of trilateration.

The maths involved can be a bit tricky to get your head around, especially in three dimensions. To make it easier, get rid of one dimension for now and look at a situation on a flat surface. Imagine a field with three trees growing around its edge. There is a cow sat in the field. If, for some reason that

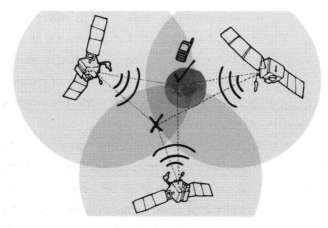

How GPS satellites and trilateration let a phone
know where it is

we need not go into, you want to plot on a map where the
cow is, you can use trilateration to do this. Start by measuring
the distance from the first tree to the cow. Now use a pair of
compasses to draw a circle around this tree on the map, and
the diameter is an appropriately scaled-down version of the
first measurement. The cow must be sitting somewhere on this
circle. Repeat this again using the second tree, and you will
have two circles on your map that cross at two points, at either
of which the cow could be sitting. Finally, using the third tree,
draw a third circle and you can then be certain that the cow is
sitting at the one point where the three circles intersect.

Your phone does all of this mathematically, without resorting
to tape measures, cows or compasses. What's more, since this
happens in three dimensions, it's not circles but spheres that the
maths has to cope with. One consequence of this is that the
phone ends up with not one, but two possible locations based

on this trilateration. Fortunately only one of these locations will be on the surface of the Earth, the other is somewhere up in space. Your GPS system sensibly chooses to ignore this second location. Another consequence is that the location generated is also in three dimensions and can give information about your height, as well as your position on a flat map.

It's a pretty nifty bit of calculation, especially when you only start with the time that the satellites transmitted their messages. However, unless you take into account the theory of relativity, GPS becomes hopelessly inaccurate.

According to the theory of special relativity, the faster you are moving, the slower time will pass relative to somebody watching who isn't moving. For things in our everyday life this isn't normally a problem, but GPS satellites whizz around the earth at about 14,000 km an hour (8,700 mph). At this speed, the atomic clocks on board will get slower by 7 microseconds a day. On top of this, you need to account for the general theory of relativity as well, which says that as the force of gravity diminishes, time will pass relatively more quickly. Since GPS satellites orbit at a height of 20,000 km (12,500 miles), the Earth's gravitational force is lower and this makes the atomic clocks speed up by 45 microseconds per day. Overall the net effect is a quickening of the clocks by 38 millionths of a second each and every day.

Now, this may seem like a stupidly small amount of time to be worrying about. However, if you plug this into the trilateration calculations done by the GPS microprocessor in your phone, it will turn into an 11-km (7-mile) error. What's more if you don't continually correct for this, that error will grow by 11 km each day. At the end of a week your GPS will be giving a location that is 80 km (50 miles) away from your actual location. Fortunately, the builders of GPS satellites are well aware of Einstein's theories of relativity and niftily use them to correct the atomic clocks on board satellites to take account of the daily 38 microseconds increase.

Being able to find out where you are on the surface of the planet is something that our society has struggled with for centuries. Huge cash prizes have been offered and vast efforts been ploughed into solving this problem. The global positioning system finally gives us a way of doing this to incredible accuracy, but only with an understanding of one of the twentieth century's two pillars of physics. I struggle to get my head round the ideas in special and general relativity, and I know I'm not alone in that. So, it's good to know that even if I'm a bit hazy on the details, within my phone is a GPS system that provides a brilliant, everyday proof that Einstein's theory works.

Different flavours of smoke alarm

The smoke alarm has become a fantastically widespread household object, and for good reason. Global statistics show that you are twice as likely to survive a house fire if you have fitted smoke alarms. National fire services around the world are extremely keen on smoke alarms as they not only save lives by alerting people much sooner to fires, but also allow the fire services to be contacted quickly, enabling them to get to the fire before it becomes too big.

While we're all glad to have the alarms, the wonderful engineering inside them is often overlooked. There are two main types of smoke alarm and they are both ideal for slightly different types of fire.

If a fire begins to burn material that is not particularly flammable, it will create what is known as a smouldering fire. This will produce smoke that consists of large soot particles, although it's all relative, and even these are only about one-hundredth of a millimetre across. To detect these sorts of particles you want an optical detector. Inside the shell of one of these, as well as the battery, siren and assorted electronics, you will find a small, black-painted chamber. At one end of this chamber is a light source that shines into the chamber. With modern alarms this is usually some sort of light emitting diode, or LED, and frequently it emits invisible, infrared light. Also inside this chamber, but not pointing directly at the light source, is a photodiode. This is essentially an LED in reverse; when light shines on a photodiode, a tiny electric current is produced. You can think of it as a very small section of a solar panel. Since the LED does not point at the photodiode, and since light travels in straight lines, no light reaches the photodiode, no current flows and the alarm stays silent. Now introduce some of our large soot particles from our smouldering fire. These rise up to the ceiling of the room on hot air currents and can get into the detection chamber through holes around the edge. The light from the LED hits these soot particles and instead of pointlessly hitting the wall of the chamber, the light is reflected by the soot. This bounced light is scattered in all directions and some of it will hit the photodiode. A tiny current is produced by the photodiode, the alarm detects this and sets off the siren.

If your fire is not the smouldering variety, but is instead the fast burning type with leaping flames and much smaller soot particles, about a thousand times smaller, you need a different type of detector. To detect such small particles you need an ionization smoke detector. At the heart of this type of detector is a truly remarkable substance: a tiny sample of americium, an entirely manufactured element. Americium is a radioactive element first created in 1944 at the University of California in Berkeley, USA. Its presence in smoke alarms sometimes gives people who fear its radioactivity cause for alarm. However, the weight of americium in the detector is usually only a third of a microgram, which is an incredibly small amount. To put this number in perspective, this is a thousand times lighter than a single grain of salt, and by this I mean the really small, free-flowing salt you get in a salt-shaker. The americium inside a smoke detector is embedded in a steel pellet and shielded in a metal chamber. The type of radiation produced by americium is called alpha particles, which are massive in size compared to other types of radiation. Alpha particles are completely stopped by even a thin metal shield. Safely within the confines of the detector, the alpha particles shoot between two metal plates. When the alpha particles hit molecules of gas in the air between the plates, they knock off electrons and create electrically charged particles called ions – hence the name, ionization smoke detector. Because these particles are charged, they allow a teeny, tiny current to flow between the two metal plates. When smoke particles enter this chamber – and it does not matter what size they are – the ions, produced by the alpha particle collisions, stick to the smoke and the current stops flowing. When the current stops, the alarm goes off.

So, which type of alarm should you have in your home? Both types of alarm will detect a fire, but each will detect a different type of fire sooner than the other. In some countries ionization alarms are not recommended or even outright prohibited. However, they are less likely to be set off by such everyday incidents as burnt toast and steam. Clearly, any smoke alarm is better than none, and where the option exists, give your fire service a call for advice. One thing that is worth mentioning is that it is estimated that a third of all smoke alarms fitted in houses are non-functional, either because of dead batteries or because they are gunged up with dust or even painted over. Having smoke alarms installed is only the start; you do need to maintain and test them regularly, too.

The vanishing transistor and Moore's law

In the spring of 2005, the Intel Corporation, the world's largest manufacturer of semiconductors, posted a $10,000 reward on eBay for a copy of the April 1965 edition of *Electronics Magazine*. On the other side of the Atlantic, self-confessed hoarder David Clark saw the posting and realized that he might just be in luck. Squirrelled away under his floorboards, for just such an event, was a collection of pristine *Electronics Magazine*s, including the much-valued April 1965 edition. David Clark was duly awarded the sum of $10,000 or £5,281 at that time.

Why the silicon chip giant Intel wanted to get hold of a forty-year-old copy of a trade journal becomes clear if you look on pages 114 to 117. Contained therein is an article by Gordon Moore that attempted to predict the future of the growth of the electronics industry, an industry that had only really begun in 1947 with the invention of the basic building block of the silicon chip, the transistor. Moore had noticed that up until that point the number of transistors that can be crammed onto a single chip had doubled every two years. He went on to suggest that this would continue for the foreseeable future. His observation became known as Moore's law and has remained true, pretty much, to this day. Three years after he wrote the article, Gordon Moore became one of the founders of the Intel Corporation. Thirty-seven years later it turned out that the industrial giant didn't have a copy of the classic article in its archives.

Since Moore's prediction, his law has proven to be incredibly reliable. From the early days of electronics when Moore was writing, through to the blossoming of the personal computer industry in the eighties, the transistor count and consequently the computing power has doubled every two years. In 1978 we thought we had reached sublime heights of computer chip design with the release of the Intel 8086 that contained over 20,000 individual transistors. Some 17 doublings later, as predicted by Moore, the latest microprocessors have the mind-boggling number of 2.5 billion transistors on them. That's gone from a two followed by four zeroes (20,000) to a two followed by nine zeroes (2,000,000,000).

It's astonishing that as the numbers have grown so much they have stuck so closely to the predictions of Moore's law, although there may be a hint of a self-fulfilling prophecy here. Since 2000, the *International Technology Roadmap for Semiconductors* has been published by a group of industry associations. This document lays out targets for the semiconductor industry, including things such as the number of transistors on a microprocessor, and it partly relies on Moore's law for its goal setting.

Unfortunately, Moore's law will inevitably be proven wrong, as even the man himself said: 'It can't continue forever, the nature of exponentials is that you push them out and eventually disaster happens.' While I'm not exactly sure it's going to be a disaster, he's right about exponentials: the rate at which an exponential grows gets bigger and bigger all the time. As the number of transistors predicted by Moore's law gets bigger, the size of those transistors has to get smaller and smaller. Eventually we will reach the point where the transistor is smaller than an individual atom, which is clearly impossible.

Well, yes and no. We are already getting to this level of miniaturization but it looks like there may be alternative, more cunning approaches to the problem. It's not just transistors that have to fit onto a silicon chip, you also need to squeeze in all the connections. One of the great leaps forward with microprocessor design has been the invention of new ways of joining up the transistors that takes up less space, leaving room for yet more transistors. Researchers have also come up with ways to get more work done by the transistors that they do have.

How long Moore's law continues to be true remains to be seen. Some analysts think we have already passed this point and in the next few years we will see the increase in transistor numbers plateau. Others, including Moore himself, think we have a bit longer yet; maybe twenty years before we see reality and Moore's law diverge. However, in the history of computing, Moore's law has faced other, seemingly unsurmountable obstacles in its path. Each time this has happened, a new approach or ingenious technology has kept us doubling transistor numbers every two years. Moore once described his eponymous law as 'a violation of Murphy's law', which predicts that if something can go wrong, it will go wrong. We may yet stay on track with Moore's law, and as he said himself, 'everything gets better and better'.

Wobbly crystals in your clock

Do you know what time it is? Chances are that if you check a clock of some sort – and we can pause while you do – you will have looked at a device that uses a quartz crystal. Many clocks and watches proclaim, in tiny letters on the face, that they are quartz clocks. But there will be no visible evidence of it, and even if you start pulling the timepiece apart you will be hard pushed to find any quartz at all.

Quartz is an extremely common substance: it's the second most abundant mineral in the world. Every time you see, or better, walk on, a sandy beach, most of that sand is made of quartz. Quartz is made of atoms of silicon and oxygen knitted together into a crystal. It has many desirable properties: it's very hard, it's transparent, it can be artificially manufactured, and it displays an unusual effect called piezoelectricity.

In 1880, Pierre Curie, long before he became the husband of Marie Sklodowska, discovered that if you squeeze a crystal of quartz it produces a tiny electric current. This was dubbed the piezoelectric effect and a year later Pierre Curie showed that it also worked in reverse. If you apply a current to a quartz crystal, its shape will deform ever so slightly. When you turn off the current, the crystal will return to its original shape and produce a little burst of electricity as it does so. Wind forward about thirty years and researchers at the Bell Telephone laboratories realized that if you made tiny tuning forks out of quartz, you could get them to resonate by feeding them pulses of electricity.

The vibrating quartz crystal that keeps time

When an object resonates, it is vibrating at what is known as its resonant frequency. Imagine a suitable volunteer child sat on a swing. The swing moves backwards and forwards about once every 2 or 3 seconds. This is the swing's resonant frequency, and if you want to drive the swing higher and higher, you need to push at this frequency. If you try to push the swing more often, at a higher frequency, it will not work, as the swing can only be pushed at its resonant frequency. Every object has its own peculiar resonant frequency that is determined by its physical properties: in the case of the swing, this is the length of its ropes or chains.

Inside a quartz clock is a tiny metal component a few millimetres (about 1/16 in) across. Inside this is a sliver of quartz crystal, usually circular these days, rather than tuning-fork shaped. When pulses of electricity are fed into this crystal it begins to vibrate, and these vibrations will be strongest at its

103

resonant frequency. After each pulse of vibration, the crystal relaxes and emits a tiny pulse of electricity. If you use the frequency of electrical pulses coming out of the crystal to time the pulses going in, you will drive the quartz to vibrate strongly at precisely its resonant frequency. The quartz is shaped by laser cutters, so that it vibrates exactly 32,768 times a second, and when I say exactly, I mean to within a thousandth of a vibration per second.

The reason that the number 32,768 vibrations a second is chosen is partly because it's in the range of frequencies within which it is easy to get quartz to vibrate but also, more importantly, if you divide this number by two, fifteen times, you end up with 1 vibration per second. Alongside the clever electronics that set up the resonance in the crystal is a second circuit that counts the pulses of electricity coming out of the quartz. Using repeated division by two, the circuit can produce an electrical pulse exactly once per second. From here it is just a matter of a tiny motor and simple gears to convert this pulse into the movement of the hands on a clock.

You may be thinking that this is all well and good, but surely even this is old technology. Computers and smartphones automatically know the time by downloading it from the internet. While this is true, they still need to keep time and they will do this even without the internet. All our modern clocks and time-telling devices contain what is known as a real-time clock circuit, and inside this, vibrating at 32,768 times a second, is a tiny quartz crystal.

When batteries die

In the year 1800, the electric battery was invented. The man who made this breakthrough was Alessandro Volta, a somewhat shy Italian whose name eventually became immortalized as the unit of electrical energy. Until that moment in history our understanding of things electrical extended only so far as momentary sparks of static electricity. Then along came Volta with his pile of copper and zinc discs separated by paper soaked in sulphuric acid that could produce a constant flow of electricity. Each set, or cell, of copper and zinc discs produced about ¾ of a volt, although they had no way of measuring this at the time, and certainly no concept of volts. Stack enough of these cells together into a battery and you can produce some seriously large voltages and start doing interesting electrical experiments, which is exactly what the scientific community promptly did.

There was and still is, however, a problem – all batteries eventually die. I realize that this isn't much of a surprise, as the batteries Volta invented are now present in devices all over our houses and are an essential part of everyday life. Even the batteries we can recharge eventually give up the ghost.

The key to understanding why this happens is to appreciate that a battery is a promise of energy held in chemical form. Inside all batteries are two different, and usually solid, chemicals linked together by a third, liquid, one. In Volta's pile he used solid copper and zinc connected by sulphuric acid, but there are innumerable other combinations. Irrespective

105

of the actual chemicals, the chemistry that takes place is the same. At one side of the cell the chemistry releases electrons, which then travel through the connecting liquid to the other side where they accumulate. The basic effect is that simple.

What is so clever about the battery is that this chemical reaction only takes place when you attach the battery to an electrical circuit. When the battery is removed from a circuit, the chemical reaction comes to a grinding halt as soon as the electrons begin to pile up at one side of each cell. Since they have nowhere to go, their presence stops the chemistry happening. The chemical energy stored in the battery stays there until you attach it to a closed electric circuit. Now the pile up of electrons can move away through the circuit powering whatever is attached. As this happens, the logjam is released and the chemistry starts once more.

A consequence of this production of electrons through chemistry is that the starting materials get used up. When these chemicals run out the battery dies. The promise of electricity that it contained has been given and it is spent.

The most common type of disposable battery these days is the alkaline battery. Rather than zinc, copper and sulphuric acid, these batteries contain zinc powder, manganese dioxide and potassium hydroxide as the connecting liquid. It is this last chemical, the highly alkaline potassium hydroxide that gives the battery its name. As the chemistry progresses within the alkaline battery, the zinc powder is turned to zinc oxide and the manganese dioxide becomes dimanganese trioxide. When most of the zinc and manganese dioxide have made this change, the chemistry dwindles and the battery dies. But it needn't be a permanent death.

If you want to recharge a battery you have to undo this chemistry, return the chemicals to their starting state and restore the promise of electricity. The theory behind this is ridiculously simple. Since all chemistry is reversible, you just need to force electricity to flow backwards through a battery and the chemistry runs backward too. While it is possible to recharge ordinary alkaline batteries, it's not recommended for a couple of reasons. As the zinc oxide is turned back into zinc, it can form crystals of metal in the wrong places and rupture the lining between the zinc and the manganese dioxide. If this happens, all sorts of new reactions can then take place, some of which produce hydrogen gas. Since the canister within which the battery is enclosed is gas tight, the hydrogen can make the battery explode, spraying its contents, including the highly caustic potassium hydroxide, around. This is why we don't recharge regular alkaline batteries.

For truly rechargeable batteries you need to use more complicated chemistry and have a more convoluted internal structure. This ensures that when the chemistry is reversed, it goes back to where it started without damaging the battery. Or, at least, it goes mostly back where it started. While reversing the chemistry in rechargeable batteries doesn't cause damage, it's not 100 per cent effective, and even rechargeable batteries have a limited capacity for rejuvenation.

The use of the word battery, with reference to an electrical device, was coined in 1748 by Benjamin Franklin, the great

American founding father and scientist. Before this, a battery was an array of artillery cannon, but Franklin used the word to describe the culmination of a party he threw at his home in Philadelphia. The party included assorted sparky experiments, a roast turkey killed by electrocution, electrified goblets of wine and glasses of flaming spirits ignited by sparks. All of which culminated in the discharge of guns from the electrical battery. The battery in this case was a series of static electricity containers, this being pre-Volta and his pile. Once 1800 rolled around and Volta came on the scene, the word was quickly adopted to describe his invention of stacks of many individual electrical cells.

For the deeply pedantic this has an important implication. In contrast to Volta's pile, each little modern battery tube contains only one electrical cell. If there is only one cell then your batteries are not really batteries at all. As with all pedantry, this subtlety should only be brought out for discussion on select occasions, and I suggest you focus instead on the idea that each battery, or cell, contains the potential of electricity held as a chemical promise.

Bursting your bubble

It's a cruel person that will burst anyone's bubble, let alone a child's. The merest touch from a finger will pop a soap bubble and they are in some sense the epitome of fragility. So much so that they are quite capable of bursting themselves, without apparent external prodding. Yet they can also be incredibly long-lived, as anyone who has watched bubbles floating on a gentle breeze can attest. These are not contradictory observations, once you delve into the science of bubbles.

The key to blowing a bubble, as any child will tell you, is to have plenty of washing-up liquid in your water. The important ingredient here is detergent molecules (see page 134), with the peculiar property that one end of the molecule is water-attracting and the other water-repelling. Detergent molecules, when mixed with water, can form what are known as soap films. These consist of a sheet made up of two layers of detergent, with water sandwiched in between. Both of these detergent layers are usually only one molecule thick, but packed tightly with all the molecules pointing in the same direction. The detergent molecules arrange themselves so that the water-loving bits are all facing the water inside the sandwich, and the water-hating bits poking out into the air around the soap film. If you curl a soap film around on itself you end up with a spherical bubble.

It is also worth mentioning that it's this peculiar layer sandwich that gives rise to the rainbow of colours you see in bubbles. This sandwich of detergent-water-detergent can be incredibly thin – easily less than 100 nanometres: that's a

109

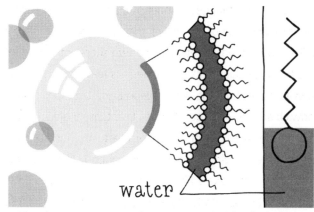

The detergent–water–detergent sandwich of a bubble

ten-thousandth of a millimetre. Now, this is a significant number because it's less than the wavelength of visible light. When light hits a soap bubble, most of it goes straight through, but some bounces off the front detergent surface of the soap film and some reflects from the second detergent layer in the sandwich (see page 214). This creates two reflected beams of light, one shifted ever so slightly out of phase with the other. Since light behaves like a wave, it's possible for two light waves to cancel each other out. This is exactly what happens in soap films. At particular thicknesses of soap film, different wavelengths or colours of light will be cancelled out. Rather than seeing white light reflected, you now see white light minus the cancelled colour. As an example, if you have a soap film that is about 430 nanometres thick, this is just right for cancelling yellow light, and the reflected light seen in the soap bubble looks blue.

The reason that the colours in soap bubbles constantly change is partly because the bubble changes shape and partly because the soap film gets thinner over time. And this is the key to what makes bubbles burst of their own accord – they dry up. The layer of water between the two sheets of detergent is just a few molecules thick, and unless it's really cold, that water will evaporate. Once the water evaporates, it escapes from the sandwich and the layer gets thinner. Eventually there is no water left, and the two films of detergent meet. At which point you have a sandwich with no filling, which nobody likes. Detergent on its own can't make a film; it needs a water filling to do this. So, the film breaks and forms a tiny hole. Once you have a hole in a soap film, the surface tension of the water inside inexorably pulls the hole bigger, and the whole film collapses.

Logically, then, if you could stop evaporation you could stop the bubble from popping. The obvious way to do this is to place your bubble somewhere with 100 per cent humidity. With such high humidity you should get no evaporation. This particular trick was mastered by the great American bubble showman with the unlikely name of Eiffel Plasterer. He blew bubbles inside large jam jars that had some water sloshing in the bottom. This water ensured a humidity as high as possible and Plasterer's record bubble lasted an astonishing 340 days.

There's also a chemical trick you can use to help stop bubbles popping. Humectants are a group of molecules that hold onto water and stop it from evaporating. You can use sugar syrup for this, but then your bubbles will leave a sticky mess wherever they land. Just as easy to get hold of is a clear, gloopy liquid called glycerol, or glycerine, as it used to be called. You can buy little bottles of this stuff in supermarkets, as it's

111

commonly added to food to keep it moist. In particular, it's often mixed into cake icing, or frosting, to stop it from setting too hard. Glycerol is usually added to the bubble mixture you buy in shops, and it really elongates the length of time the bubbles will last. From my own experience, for a good bubble mix you need 1 part glycerol to 10 parts washing-up liquid, made up to 100 parts with clean water. Mix it up really well and then leave it overnight for the froth to settle down, as the one thing that gets in the way of making good big bubbles is lots of other bubbles. Give it a try, and you should be able to blow big and long-lived bubbles.

In my career as a science communicator I have on several occasions performed with bubbles on stages and in front of cameras. At one time I was a dab hand at blowing bubble sculptures, and I even nearly broke the world record for the largest indoor bubble. Sadly, I was short by a fraction of a cubic metre. If my experience in the bubble field has taught me one thing, though, it's that the most likely way for a big bubble to burst is not evaporation, as bubbles rarely get a chance for this to happen in front of an audience. Usually, the cause of breakage is when a child pokes their finger into it. Although, now I think on it, it is not only children that can't keep their fingers away. Irrespective of the age of the finger that is the culprit to this bubble destruction, when it touches the sandwich of detergent and water it presses the water out. When this happens, you are back to the same situation as when the water evaporates. No water, no sandwich, and even if I didn't do it, I'm sad to say your bubble has been burst.

Bottled clothing

Back in 1979 the Malden Mills company in the USA released a new type of fabric they called Polar Fleece, that was intended to mimic and improve upon woollen fabrics. These days you can buy Polar Fleece in any clothing shop, although since the owner of Malden Mills, Aaron Feuerstein, chose not to patent the invention it's usually just called fleece these days. What would probably have surprised Mr Feuerstein back in 1979 is that today vast quantities of fleece are produced, and most of it is made from recycled plastic bottles.

The Polar Fleece that Malden Mills created was woven from unrecycled and entirely synthetic polyester threads, which was not a new method, but what made it special was what they did to the cloth after it had been woven. They applied a couple of established cloth-making techniques to the fabric. Firstly, the cloth was brushed with fine wire combs, to pull little loops of fibre up from the surface of the cloth in a process called napping. The tops of these loops were then sliced off to create a fuzzy fabric. Polar Fleece was initially just a hit with backpackers and hikers, but is now a popular part of everyday clothing. Few among us do not have some sort of fleece tucked away for those cold winter days.

Back when Malden Mills first started to make fleece, the raw ingredient for the polyester fabric was a plastic with the unpronounceable name of polyethylene terephthalate, or PET, if you decide to chicken out of the full name. At the time, all the PET being used was made from chemicals derived from crude oil. However, it turns out that PET plastic is not just good for making clothing fibres. It is also perfect for moulding bottles: the global bottled drinks industry relies on PET. What makes PET such a useful material is that it's a thermoplastic, which is just a posh way to say that when you heat it up it turns back into a runny liquid. If you take an empty PET bottle, it melts when you heat it to 250 °C (482 °F). Now squirt this liquid through a heated spinneret, which is essentially just a fancy showerhead, and you will create polyester fibres. It sounds relatively simple, but it turns out to be much more fiddly in reality, and it's a testament to the recycling industry that we can achieve this.

The collection and recycling of our used PET bottles is now something that we take for granted, although it's clearly a significant logistical task. Once gathered up, these bottles have a long journey ahead of them before we end up with a new fleece – quite literally in many cases. PET bottles are usually hand-sorted to remove any large undesirable items, such as twigs, or cans that have been incorrectly recycled. The baled-up bottles are then, in almost all cases, shipped out to the Far East. Here they are shredded, and as the caps are made from a different plastic – high-density polyethylene – these are separated using water. Despite its name, high-density polyethylene floats on water, while PET sinks. Removal of paper labels and the glue used is a further, significant problem that needs the application of some unpleasant chemicals, such as caustic soda. After this process, you have clean shreds of wet PET bottle, and the water is an issue. When you melt PET, any water present will be incorporated into the plastic, which causes it to start degrading and breaking down. One of the trickier problems for PET recyclers is drying out all their shreds in an energy efficient way, without starting the melting process. Finally, the PET shreds are ready to be made into fibres, but even this is quite a palaver.

The shreds are melted down and squirted through spinnerets to make fibres. However, since the fibres aren't quite thin enough at this point they are reheated and stretched before being crimped and chopped up into wiggly strands about 4 cm (1½ in) long. The resulting fluff can be used to fill pillows and cuddly toys, or for fabrics it is combed and spun into thread. At last you have the wherewithal to make a polyester fabric and begin the process invented in Malden Mills, but unlike in 1979, this fleece will have been made from recycled bottles rather than crude oil.

The non-shrinking sheep

Wool is a remarkable material. It can be used to create clothes that are some of the warmest, most robust, most wrinkle free and even most breathable. Sure, the best of high tech fabrics can out-compete woollens in one or two categories, but if you want an all-purpose, natural fabric, it can't be beaten. However, despite all this approbation, it does have one major flaw: it shrinks if you don't wash it carefully. I have been the instigator of a number of miniaturized sweaters, and in the aftermath of each such event I have always wondered why it is that sheep don't have the same problem. After all, in my neck of the woods it rains aplenty, yet I have never seen a sheep with an uncomfortable and slightly too small fleece.

Wool is composed primarily of a protein called keratin that forms long coiling strands which are themselves coiled and bundled together and set into a matrix. The whole strand is coated in a final layer of dried-up cells that are packed with more keratin. It's essentially the same structure as human hair and animal fur. In fact, and pardon the slight diversion, the only difference between hair, wool and fur is the density of strands growing from the skin. An average human head, not suffering from male-pattern baldness, has about 40 hairs per square centimetre (260 per square inch). Compare this to a Merino sheep that has 9,000 stands of wool per square centimetre (58,000 per square inch), and the record for fur, held by the sea otter with over 120,000 strands per square centimetre (775,000 per square inch). Clearly, while the basic internal structure may remain the same, the more densely packed wool and fur tend to have much finer strands.

One of the reasons that wool is so suited to being spun, and made into cloth, is that it has both crimp and elasticity. The latter quality is exactly as it sounds: wool fibres can be stretched and they will then spring back to their original shape. Partly, this is to do with all the coiled strands of keratin protein that make up the wool, but it is also to do with the crimp. Wool is naturally wavy and has anywhere between 1 and 12

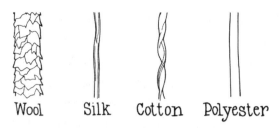

Scaly wool fibres and smoother alternatives

kinks in a single centimetre (2 to 30 per inch). The wool strands are actually shaped like stretched springs. This coiling comes from the internal structure of the strand, one side being slightly stronger than the other. Since this difference is integral to the coil, a wool fibre always returns to its coiled shape, no matter how many times it's stretched out. Together, the crimp and the keratin make wool a stretchy fibre.

We now turn to the dried keratin scales found on the surface of the wool fibres. All hair, wool and fur fibres have a coat made of overlapping dried-up cells. As you might have seen in hair care commercials, they are a bit like the tiles on a roof, but in wool the scales don't lie particularly flat. The edges stick up a bit and tend to get caught on one another. And now we finally get to the reason that woollen clothes shrink.

If you place wool in hot, soapy water, the fibres absorb some of the water, and swell a little bit. You don't actually need the water to be hot or soapy, but the swelling is much more pronounced if it is. As the wool swells, the edges of the scales push up even further than usual. If you begin to agitate the wool, perhaps by swirling it back and forth in a washing machine, the fibres ratchet together as the scales lock onto each other. What's more, the hot water relaxes the spring in the fibres a little bit. This allows the scales to pull the fibres together with less elastic resistance. When the wool subsequently dries and cools, the spring returns and the fibres tighten up, locked together by their own scales. You have now successfully shrunk your sweater. In fact you are partly on the way to making felt, which is just wool that has been heated and pounded until all the fibres are tightly locked and held together by the spring of the wool. While felt is a valuable and useful material in its own right, it's much less flexible and comfortable to wear, so this is not something you want to do to your sweater.

Since wool can shrink even in cold water, this science would indicate that sheep should suffer the same fate in cold rain. But they don't, because there is another component of a sheep's fleece to take into account.

Lanolin is a yellow, waxy substance secreted by the skin of sheep that can account for up to a quarter of the weight of a single fleece. It's a substance analogous to the sebum secreted by our skin that makes hair gradually become greasy (see page 134). Lanolin not only gives the wool its unique sheepy smell, but also provides a waterproof coating. The lanolin prevents the fleece from becoming soaked with water so that the wool never gets particularly wet. What's more, waxy lanolin covers the keratin scales on the surface of the wool strands, so that the wool can't lock itself together. It's lanolin that prevents the wool on a sheep's back from shrinking.

In theory you could make your woollen garments shrink-proof by drenching them in lanolin, but I fear, even if you could put up with the smell, that those around you would object. Fortunately, technology has come to the rescue with machine-washable wool for some garments, although please do check the label before you chuck your favourite sweater in the washing machine. One way to achieve this involves plunging the wool briefly into a bath full of acid, dissolving the scales on the surface of the strands, so the wool can be machine washed. Alternatively, the wool can be coated in a lacquer that covers up the scales. Either way, by mimicking what sheep do with lanolin, the result is a fibre that has all the properties of wool, except its propensity to shrink.

Fresh air really is good for you

There is a deep-rooted idea in the Western world that fresh air and sunshine are good for us. Going back to medieval times, the spread of disease was often blamed on bad or malodorous air. The idea that fresh air could cure was current right up until our recent history. During the nineteenth century, medical health professionals such as Florence Nightingale encouraged patients to take fresh air, and the only treatment for tuberculosis was a stay at a sanatorium. Medical care at such places, which lasted right into the early twentieth century, consisted of the current most popular healthy dietary regime and being plonked outside in a chair, wrapped in blankets come rain or shine. Fresh air and sunshine were regarded as the cure for all that ails.

We now look back on these notions, armed with our thoroughly modern medical ideas and smile knowingly at the quaintness of such a concept. While we understand that sunshine is essential for manufacture of vitamin D within our bodies, the idea that sunshine and fresh air are part of a hospital treatment seems backward. It may be psychologically helpful having the sun shining and a gentle breeze blowing through a window, but surely it can't have a direct medical effect?

Well, it looks as though all those medics of yesteryear may have been on to something after all. In the case of sunshine, we now know that ultraviolet light is rather effective at killing bacteria. In particular, ultraviolet light at the specific wavelength of 207 nanometres is fully absorbed by teeny bacteria, but causes minimal damage to much larger human cells. All that absorbed ultraviolet energy causes damage to bacterial DNA and kills them stone dead.

More intriguing is the effect of fresh air in hospitals. Hints that unfiltered outside air may have a beneficial effect have been around for a few years now. In a study of American soldiers on duty in Saudi Arabia during the first Gulf War of 1990, it was found that people sleeping in tents suffered fewer colds and sniffles than those in air-conditioned accommodation. You may imagine that the recycling of air in the air-conditioned dormitories was responsible for its greater rates of contagion. However, air-conditioning rarely recycles air; instead, it filters and then cools external fresh air.

Then in 2012, Professor Jessica Green at the University of

Oregon in the USA published the results of bacterial samples taken from surfaces in hospital rooms. Some of the samples came from air-conditioned rooms while others came from rooms with regularly opened windows. What Professor Green found was that while the number of bacteria present did not vary so much, the type of bacteria did. In air-conditioned rooms there were fewer types of bacteria, but many of those present were potentially disease-causing in humans. If you routinely scrub a hospital room with soap and anti-bacterial agents, you will kill off most of the bacteria in the room. However, it is a constant battle, and the room will immediately be recolonized by bacteria from the most abundant source, which of course is the people in the hospital, many of whom are poorly patients harbouring pathogenic bacteria. So, it's not really surprising that hospitals are full of nasty bacteria. What is more surprising is that if you open a window, many of these nasty bacteria are not present.

It is only recently that we have begun to appreciate how rich and diverse and ubiquitous the bacterial ecosystem that surrounds us really is. Unfiltered, un-air-conditioned, fresh air is full of a huge variety of bacteria floating about on dust and in tiny water droplets. If a surface is exposed to this air, then these bacteria will settle and where suitable resources are available they will grow and flourish. Most of the bacteria are harmless. Those few bacteria that potentially cause disease or infection have to compete with a whole range of other bacteria; consequently, they cannot predominate and are less of a risk to us.

If you open a window in a hospital room, whose surfaces

are regularly disinfected, those surfaces will be recolonized by a vibrant ecosystem of bacteria containing far fewer of the nasty variety. This seems to imply that we should stop cleaning hospital rooms and just open the window, but it is not that simple. Regularly cleaned hospital rooms still have fewer bacteria overall, which is a good thing when vulnerable people are using the room. But if the windows are opened between the cleaning, the bacteria that are present will be less likely to cause a disease or an infection.

Florence Nightingale became a household name after her brilliant nursing work during the Crimean War. One of her most significant innovations was the rigorous application of cleanliness within her hospital wards, including a constant supply of fresh air. On her return to the United Kingdom she carried on her pioneering work, and is largely credited with the founding of the modern nursing profession. While many of her principles have remained in current practice, her ideas on fresh air have fallen out of vogue. Perhaps it is time to revisit this idea. As she put it, over a hundred years ago in 1898: 'Never be afraid to open windows.'

The Peculiar Human and the Science of Us

Totally tasteless myths

There are two things that we have all been taught about taste: that there are four different tastes we can detect, and that the tongue is divided into distinct regions where these can be detected. The four tastes are bitter, salty, sweet and sour. The tongue detects sweet at the very tip, salty at each side, sour further back, and bitter as a band across the very back. If you look in textbooks, websites and home science books you will find these facts confidently presented. And yet, these ideas are completely wrong. There is no taste map on your tongue, and there are more than four basic tastes.

The myth of the tongue map should be easily dismissed. If you blindfold a test subject and place dots of flavoured liquid around the tongue, you can plot where the different tastes are found. When carefully tested, each taste can be detected all over the surface of the tongue. Yet it is precisely this experiment that has been performed by countless school children across the world to prove that a taste map exists. It's a wonderful example of how prior knowledge of an expected result can bias your findings.

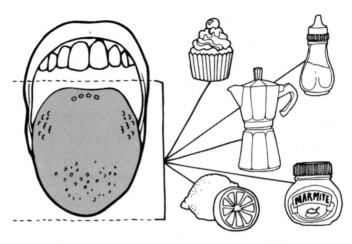

**There is no taste map on the tongue, and there
are at least five basic tastes**

Presumably all those school kids either fudge their results to give the classic taste map, or their correct mapping is dismissed as poor experimentation. If you trace the introduction of the taste map in scientific literature, you find its origin in a poorly written German paper mis-translated by an American scientist in 1901. It's a persistent scientific myth, but probably not as ingrained as the idea that there are only four basic tastes.

In all likelihood, the Ancient Greek philosopher Democritus first catalogued the different tastes as part of his groundbreaking atomic theory. Democritus came up with the idea that everything is made of minuscule, indivisible blobs of stuff he called atoms, and the taste of an atom was determined by its shape. So, sweet things were smooth blobs, salt had sharp edges, sour was bulky and angular and bitter things had jagged hooks. The various ideas that Democritus had about taste finally settled into those four tastes, and we believed that for thousands of years. Then, at the start of the twentieth century, people began to realize that there was more going on. One of these people was Auguste Escoffier, who was dubbed by the French as 'king of chefs and chef of kings'. While Escoffier was busy revolutionizing cooking and the restaurant experience in Paris, he was also coming up with new recipes that tasted neither sweet, bitter, sour nor salty. Since, according to science, there were only the four tastes, Escoffier was producing something impossible and magical.

At the same time, in Japan, a brilliant chemist called Kikunae Ikeda had an epiphany while eating cucumber soup. On the day in question in 1908, his soup was more delicious than usual and he realized that this was due to the addition of kelp seaweed. It took him six months to isolate the chemical in kelp that was responsible for this transformation, and what he found was the amino acid known as glutamate. Amino acids are the building blocks of protein and glutamate is common in most proteins – it makes up about 6 per cent of human proteins. Initially it was assumed that glutamate was enhancing the other four tastes, and Ikeda shrewdly patented the process for manufacturing sodium glutamate as a flavour enhancer. This is of course the form we are most familiar with: monosodium glutamate. Ikeda dubbed the taste effect of glutamate as *umami*, which literally means 'delicious taste' in Japanese. Nearly one hundred years later, in the year 2000, researchers identified the receptors on your tongue that detect glutamate, at which point, umami was unquestionably declared the fifth basic taste.

It's very difficult to describe a taste as they are so fundamental to our experience of the world, but umami is usually associated with a full, savoury flavour. It's found in meat broths, anchovies, shiitake mushrooms, Marmite and hard cheeses such as Parmesan. It turns out in retrospect that cooks have been using umami-rich ingredients in dishes for centuries to create rich savoury flavours. Now that we have a chemical understanding of this new taste, it's possible to find umami in dishes where it never was before. You can create 'umami bombs' and even buy umami pastes, both vegetarian and non-vegetarian.

So, there is no taste map and there are five basic tastes. Well, except that there is a debate about a recently discovered taste receptor that can detect fat molecules. It may turn out that we can also taste fats. All of which means it looks like Democritus may have been closer to the truth all along. In his original work, written about 400 BC, he described not four but six tastes. These were the four traditional ones plus two others that he called pungent and oily. We may be only now identifying the oily taste and it's not entirely out of the question to imagine that pungent is the umami taste. I like the idea that it's taken us 2,500 years to find ourselves back where we started.

Cracking knuckles and a fifty-year experiment

I will admit to being a knuckle cracker and also a foot cracker. It's not something I flaunt in public, but I probably privately crack the joints of my hands and feet a couple of times a day. There has always been a niggling worry in my mind about this habit. The popular belief is that cracking your knuckles is bad for you in the long run, with many authorities, such as uncles, aunts and parents, telling you it will cause arthritis. It can certainly sound quite alarming when a finger joint, toe or back emits a loud crack. It doesn't feel like part of the normal workings of the human body, and surely it can't be good for you? What's more, irrespective of the possible pathological consequences, what on earth makes your joints do this?

Surprisingly, for something that happens spontaneously to all of us and many of us actively seek it out, the reason why joints make cracking noises is not fully understood. The current leading theory on this subject goes like this: in between the bones of each joint is a bag of fluid called a joint capsule. The fluid inside the bag is mostly water, but also contains some proteins, salts and white blood cells. It's there to help lubricate and cushion the ends of the bones that form the joint. When you attempt to crack a joint, you rapidly stretch out the joint, which in turn stretches the joint capsule. This drops the internal pressure of the capsule and tiny gas bubbles appear inside the fluid in a process called cavitation. The key to understanding why this happens is that a liquid cannot be compressed or stretched. So, if you have a bag completely filled with a liquid and you suddenly increase the size of the bag, the liquid cannot be stretched out to fill the new, bigger volume. Instead, bubbles appear to make up the extra space. These bubbles are usually filled with gases that were, up until the bubble formed, dissolved in the non-stretchy liquid.

In the case of a joint capsule, the liquid inside has lots of nitrogen dissolved into it and the idea is that when you crack your joints you are making bubbles of nitrogen inside the joint fluid. The noise associated with the cracking action doesn't come from the creation of the bubbles, but when these bubbles rapidly collapse. What you hear is the liquid smashing back in on itself to close up the bubble after the joint has been stretched. At least, this is the theory. Up until now, no one has managed to study cracking joints in real time, to prove that this is actually the cause of the crack.

While we haven't yet proved where the noise comes from, we do have more information about the potential rheumatic implications of cracking. In 1998 a small communication appeared in the scientific journal *Arthritis & Rheumatism*. It was authored by Donald Unger, a seventy-two-year-old medical doctor from the city of Thousand Oaks, which is just outside Los Angeles. In it he described the sort of authoritative advice he had received as a young man from an assembly of family experts. However, unlike most people, Donald Unger's reaction was not to politely acknowledge the advice and make a note to crack his knuckles in private. Instead, he set out to perform a definitive experiment, albeit with a very small sample size. He began to crack the knuckles on his left hand twice a day, while leaving his right hand un-cracked. In a display of remarkable perseverance he continued this for fifty years. At which point he examined the data he had. Neither hand showed any sign of arthritis.

Dr Unger's is not the only study to look at this; there have been two others that both sought out a correlation between cracking joints and arthritis. These studies both questioned elderly people on their joint cracking habits and noted which patients also had arthritis in the cracked joints. In both studies there did not appear to be a link between arthritis and joint cracking. While the sample sizes of these studies were significantly bigger, and thus possibly more scientifically convincing, they did not set about actively cracking knuckles in a systematic way. It should be noted that performing experiments on human beings to try to cause arthritis by cracking knuckles would certainly not get past any medical ethics committee these days.

I also noticed that the advice not to crack one's knuckles given by concerned relatives to young scientists was the impetus to not only Dr Unger's experiment but also one of the other studies. In this second example of folk wisdom, it was grandmotherly advice to the twelve-year-old son of another medical doctor that prompted the investigation. All of which makes you wonder, should we be questioning parental advice more often? As Doctor Unger put it, does this also shed doubt on his mother's exhortation that he should eat his spinach and broccoli?

Extrasensory perception

It's fairly standard practice to teach children a simplified version of the complexities of science, but there is one popular mis-teaching that pervades our culture much deeper even than primary school science classes. What's worse is that it's a concept that doesn't even make sense when you start to look closely. I'm talking about the idea that the human body has five senses with which we perceive the world around us. It's an ingrained idea, as we have all been taught that these five senses are hearing, sight, smell, taste and touch. Scratch the surface even a little bit on this subject and it's quickly apparent that we have many more ways to sense the world.

My own personal favourite counterexample to the five senses theory can be demonstrated with the following little trick. Start by positioning yourself somewhere that allows your arm to move freely, without touching anything else. Then close your eyes and stretch your arm out wide to the side. Now touch your nose with your eyes still closed. Did you manage it? For most people this is no challenge, but for those who have problems with what is known as proprioception, or sensing where your body parts are, touching your nose this way is a bit more of a problem. Some sceptics may argue that this is nothing more than the geometry of your body dictating that when you bend your arm, your finger hits your nose. In which case, try the advanced proprioception test: this time don't touch your nose, but with your eyes still closed, stop your finger just in front of your nose by a couple of centimetres (about an inch).

Try it. It feels decidedly uncanny that you can do this without any comedy poking yourself in the eye or up the nose.

Attached to every single one of our skeletal muscles is a stretch receptor that reports back to our brains how stretched out or relaxed each muscle is. As babies, we all unconsciously construct a model of our bodies in our minds and how all the different stretch receptors relate to where our limbs are. It's the reason you can tell what expression you have on your face without having to look in a mirror. If our proprioception goes wrong, when under the influence of alcohol for example, you become clumsy, knock things over and trip over your own feet. You literally don't know where your arms and legs are.

The other most obvious sense not listed in the big five that goes hand in hand with proprioception is our sense of balance. This is all down to the arrangement of tubes of fluid inside your ears that are detecting acceleration, which is to say that your ears know when you change your speed or you change the direction of your speed. The most obvious use of this is our ability to detect when our bodies are not upright and we are in the act of falling over. It's also a sense that's easy to disrupt and confuse: we are all familiar with the sensation of being dizzy, usually because you've spun around, and the fluid in your ears keeps spinning while you don't.

There are other senses too, and they all start to blur the lines on what is and is not part of a different sense. There are receptors all over the insides of our bodies that can detect the levels of different chemicals such as carbon dioxide, salt, oxygen and hormones. From a purely functional perspective these senses are the same as our ability to smell, just located in a different place. You could say that your brain can smell carbon dioxide, but clearly not with your nose. You are also not consciously aware of the results of your brain sensing carbon dioxide. Your brain automatically adjusts your breathing rate without a conscious decision on your part.

Similarly, take two of our other abilities – to detect heat and to detect pain. Both of these function in a similar way to our sense of touch, and yet at the end of the sensory nerves there are distinct microscopic structures that make them different. For that matter, our sense of touch comes from a range of different types of sensors each named after the nineteenth-century anatomist that discovered them. You have Meissner's corpuscles that respond to light pressure, Pacinian corpuscles that detect hard pressure and vibration, Merkel discs that sense prolonged pressure and Ruffini endings that perceive stretching of the skin. So, does this mean that each of these is a different sense?

When you start to examine the biology of how we perceive the world, it all gets a bit complicated and messy. Our senses don't fit neatly into five little boxes. Even something like vision, which seems fairly straightforward, turns out not to be. We have two ways to see things: one way uses rod cells and sees in black and white, in fine detail and also in low light levels. The other way of seeing relies on three types of cone cells that detect colour, but not in as much detail, and this way needs more light. The sensitivity of these visual systems turns up and down, depending on what conditions we are in. If you walk from a light space, where your cones are doing most of the work, into a dark room, your rod cells take over, and your night vision slowly kicks in, over half an hour or so. Which all seems to imply that we have two ways to perceive light and thus two senses of vision.

We clearly don't have just five senses, but a plethora of different ways to perceive the world around and inside our bodies. Some of these senses, once we start to examine them more closely, start to overlap, and the lines become blurred. The idea that we have only five senses is clearly a handy way to teach basic biology within primary schools. The danger and confusion comes when we stop using our own senses to notice that this is a simplification, and we need to embrace the somewhat less orderly and distinctly messy reality of biology.

Chemical juggling act on your hair

In 1987, Proctor & Gamble introduced *Pert Plus* onto the market, a new type of hair care product. It was a 2-in-1 shampoo that would, they claimed, both wash and condition your hair at the same time. It was the first of many 2-in-1 shampoos that have been described as 'revolutionary' or '*truly* life-changing'.

While I am pretty certain that taking only one bottle into the shower rather than two has changed nobody's life, it is an

133

interesting and perplexing product. At the time, the idea of a 2-in-1 shampoo was greeted with scorn by chemists, since the acts of shampooing and conditioning are essentially opposites of each other.

Hair gets dirty for a couple of reasons. Firstly, numerous strands of hair are perfect for picking up and trapping dust, flakes of skin and a range of environmental pollutants. Secondly, at the base of every shaft of hair is what is known as a sebaceous gland. This secretes an oily liquid that coats the hair, keeping it supple but also exacerbating the accumulation of dirt. When you shampoo your hair you are pouring a surfactant onto your head in an effort to remove all of this dirt. Surfactants are a type of molecule that include both soap and detergents, and which exhibit the peculiar property of being both water-loving and water-hating at the same time. The simplest surfactant molecules are just long chains of carbon atoms with a couple of oxygen atoms at one end. This arrangement means that the long chain of carbon is water-repelling, while the terminal oxygens are water-attracting. When a surfactant encounters oil, it surrounds teeny, tiny blobs of it with all the water-hating ends of the surfactant molecule pointing inwards to the oil. The blob of oil is now coated in surfactant with the water-loving ends pointing out. This allows the impossible to take place: you can now mix oil with water, and water will wash away the oil with the surfactant. All of which means that when water and shampoo are applied to your hair, they will strip away oils secreted by the sebaceous glands and trapped dirt, and your hair should come away squeaky clean. Unfortunately, your hair will then be bereft of oil and might become dry, possibly frizzy, and prone to becoming tangled. To alleviate these issues while keeping your hair clean, you need to reintroduce lighter and more fragrant oils back into your hair – which is what hair conditioner does. So, shampooing removes oil and conditioning puts it back. It's easy to see why the introduction of 2-in-1 shampoo was greeted with scepticism. But it does work, and its success is down to three chemicals working in concert.

The first chemical to consider in 2-in-1 shampoo is the surfactant that does the job of cleaning the oils from your hair: this is usually a laureth sulphate or a lauryl sulphate. Then, for the task of putting oil back on your hair, a chemical called dimethicone is usually required. This is an interesting chemical that consists of a long chain of silicon and oxygen molecules with carbons stuck on the outside. It's particularly good at sticking to hair, coating it and leaving it glossy and smooth. The last magic ingredient is quaternized hydroxyethyl cellulose, or polyquaternium 10 to its mates, which has two useful functions for your hair: it makes your hair less prone to static, so that it lies flat, and it partially stops the build-up of dimethicone. How it does this we don't really know. As is often the case with cosmetic products, the exact science of what is going on lags far behind the relentless drive for innovative products.

This still does not fully answer the question of how you can simultaneously add both a surfactant and an oily substance to greasy hair, strip away the natural oils and leave those fresh ones that you added. Partly, this is because the surfactants we use are better at stripping away natural oil and less good at removing the dimethicone. But it also has to do with the imperfect way all of these products work. When I say that a surfactant strips oil from your hair, what I should say is that it *mostly* strips away the oil. Any scientist will tell you that life is never so black and white. The surfactant in 2-in-1 shampoo removes most of the oil from your hair and some of the conditioning oil you are trying to add back, but leaves enough to condition your hair.

We have enjoyed 2-in-1 shampoos for over twenty-five years and they remain a popular product. However, if you look at the ingredients list on a regular bottle of shampoo you will find that the magic 2-in-1 ingredients have crept into these products as well. Most shampoos these days, no matter the variety, contain some dimethicone and polyquaternium 10, just to make sure that after you wash your hair you feel suitably pampered and glossy.

Super-strength teeth with fluoride

Every tube of toothpaste in the supermarket proudly proclaims that it contains fluoride for strong teeth and cavity protection. In many parts of the world, fluoride is routinely added to drinking water to promote healthy teeth and dentists apply fluoride whenever children attend their clinics. What's it all for, though?

A bit of basic dental biology first. A tooth is primarily made up of a hard, mineral-containing substance called dentine. Inside this is the pulp where all the nerves, blood vessels and connective tissue can be found. When you peer inside your own mouth, none of this can be seen. Instead, you see the layer of enamel that coats the dentine and makes up the rest of the tooth. Enamel is a remarkable material: it's the hardest stuff found in the human body and is the reason that we can chew on ice cubes and test gold coins by biting them. Enamel is harder than steel – on the Mohs scale of hardness, enamel is a 5, while steel is only 4.5. That said, I don't suggest trying to leave your mark in a steel girder by biting it, as the dentine underneath your teeth's surface enamel will shatter. We need enamel to be so hard because it has to last us a lifetime, quite literally.

Enamel is made up of a chemical called hydroxyapatite, which is essentially just a crystalline form of the calcium phosphate that you find in milk, cheese and dairy products. About 96 per cent of a tooth's enamel layer is hydroxyapatite, the rest being water and a tiny bit of organic material. It's usually no thicker that a couple of millimetres (about $\frac{1}{16}$ inch) and since it's just solid calcium phosphate with no blood supply, it can't be regrown if it's somehow lost. It is a wonder material with one unfortunate Achilles' heel: it's susceptible to acids.

Inside your mouth are your oral bacterial microflora – basically, a bunch of bacteria living in your mouth and feeding on whatever you eat, in particular sugar. When these bacteria – *lactobacillus* to name a common culprit – get hold of sugar they start frantically growing, dividing and producing lactic acid as a waste product of their digestion. If the acid-producing bacteria are stuck to the surface of, or nestled in a crevice on, a tooth, the acid dissolves away some of the calcium in the hydroxyapatite and the tooth becomes demineralized. When this goes on unchecked, the bacteria and acid eat a hole in the enamel and a cavity begins to form in the much softer dentine underneath.

If this were the only process involved, our teeth would dissolve before we were out of our teens. Thankfully, this is only half the story. Once you finish eating, saliva washes away most of the sugar and returns the inside of your mouth to normal levels of acidity, or neutrality, to be precise. At which point, remineralization begins to take place. All that calcium lost to the acid hops back into the enamel and repairs the damage done. That is the theory at least, but if you don't let that remineralization happen, for example by constantly swigging sugary drinks, you will suffer from tooth decay.

This is where fluoride comes into the story. Fluoride is just an electrically charged form of the chemical element known as fluorine. It's found in all manner of things that we regularly eat and drink, such as raisins, carrots, wine and meat; tea is a particularly good source. In many places, it occurs naturally in drinking water. When fluoride is present in your mouth, with low acid levels and the process of remineralization occurring, the fluoride is incorporated into the tooth enamel. The calcium phosphate snuggles up with the fluoride to become either fluorhydroxyapatite or, if it gets really cosy, fluorapatite. This has a number of benefits to you and your teeth. Firstly, fluorapatite is more resistant to acids than hydroxyapatite and once you have a layer of it on your teeth, they are significantly less likely to be demineralized and you'll get less tooth decay. Secondly, having the fluoride in the enamel helps get the calcium back into the tooth and encourages the remineralization process itself. Finally, it looks like the fluoride may get in the way of the bacteria's digestion of sugar and the production of acid in the first place. You get a triple whammy of more resistant teeth, more remineralization and less demineralization. Ideally you want fluoride hanging around in your saliva between meals, and this is why dentists and toothpaste tubes recommend that you don't rinse your mouth after brushing. This is also why putting fluoride in drinking water makes sense, so that it will tend to be present in the mouth more often than twice a day.

While the basic science of what fluoride does to your teeth would point to it being a good thing, the subject of using fluoride, especially in drinking water, is particularly controversial. If you take a peek at this subject on the Internet, you will see that it is a hot topic of, er, lively debate. It is also the subject of some truly terrible misunderstanding of science. It presents a classic example of damnation by association, where the presence of fluorine in different chemicals is taken to imply danger. Hydrofluoric acid, for example, is well known as being an unpleasant and toxic chemical, with the somewhat disturbing property that it will dissolve glass. However, the fluoride in toothpaste bears none of the properties of hydrofluoric acid.

That said, fluoride can become toxic at high concentrations, as are many of the substances we routinely encounter, but you would need to consume about fifty tubes of toothpaste in one sitting for there to be any danger. There is also another problem found in areas where there is a very high, naturally occurring fluoride level in the water. It can cause something called fluorosis, where your teeth become discoloured, or in extreme cases it causes damage to bones and joints. In areas where this can happen, fluoride is routinely removed from the drinking water to reduce it to a safe level. The levels of fluoride you are exposed to in treated drinking water and toothpaste are far too low for the effects of fluorosis to show up.

Both the dental science and the big studies of populations show that fluoride is great at reducing dental decay, especially for adults and children who can't afford top-quality dental care. There is one last factor to consider, though, and this is at the heart of much objection to putting fluoride in drinking water. Does the addition of fluoride to drinking water constitute mass medication of a population without their individual consent? This, however, is a question of politics, ethics and, ultimately, philosophy.

Prune time in the bath

Having a bath has uses aside from getting clean and reading a good book: it gives you an excuse to observe a bodily quirk that it would appear has been with us since *Homo sapiens* left the trees. Finger- and toe-wrinkling takes about five minutes of full immersion for the effect to begin, but if you want maximum, scientifically proven wrinkles, then soak in salty water at 40 °C (104 °F) for thirty minutes. If you seek out an explanation for this phenomenon, you will overwhelmingly find the following science and – spoiler alert – it's rubbish:

> The outermost layer of your skin is called the *stratum corneum*. It protects us from cuts, abrasion and general wear and tear. According to standard finger-wrinkling theory, water soaks into the *stratum corneum*, swelling the cells and making this top layer of skin expand. Immediately beneath this layer is the *stratum granulosum*, that is full of water-resistant fats. This blocks the water from soaking in any further. So, the outer layer expands and the inner layers stay the same. To accommodate this growth, the surface of your skin puckers up. Furthermore, only your fingertips and toes wrinkle, since they lack sweat glands that secrete waterproofing. Finger-wrinkling is a by-product of our biology, nothing more.

It's a simple, elegant theory that is unfortunately completely wrong in pretty much every aspect. Which is a shame, as

141

simple elegance should be encouraged. However, it turns out that the real explanation for wrinkles is far more surprising, rather useful, and may tell us something about our pre-human ancestry.

In 1936, two researchers at St Mary's Hospital in London were studying patients with paralysis in their arms. Thomas Lewis and George Pickering knew that the paralysis was caused by damage to the main nerve that runs down the arm all the way to the fingertips. What came as a surprise was that the fingers of their patients did not wrinkle when soaked in water. The patients' non-paralysed hands and their feet showed normal wrinkling, but the wrinkles were absent from the paralysed hand. Nobody took much notice at the time and the story went quiet until 1973, when an Irish plastic surgeon, Seamus O'Riain, noticed essentially the same thing – or, at least, the mother of one of his child patients noticed smooth fingers where once there had been wrinkles.

It turns out that your fingers wrinkle because your body tells them to wrinkle. It's not a passive effect, but an active decision that your body takes. We aren't conscious of making that decision because it's controlled by the automatic part of your nervous system that controls your breathing, heart rate and sweating. This can lead to some weird effects. For example, people who have severed fingers stitched back lose the wrinkles from the reattached fingers, until the nerves regrow and feeling and wrinkling return. You can also mess around with this by using drugs that shut down parts of the nervous system; they also turn off finger-wrinkling. The evidence is conclusive: finger-wrinkling is nothing to do with water soaking into the skin and everything to do with an active control, albeit one of which we are not conscious.

How, then, does our body turn on and off finger-wrinkling? This is where the science becomes a bit more fuzzy and uncertain. Your fingertips do have sweat glands, despite what the commonly believed theory states. When sitting in a bath, water runs up these sweat glands and somehow signals to the body that we are soaking in water, which is why wrinkling takes a little while to get going. How this sets off the wrinkle signal is not known. One suggestion is that the water dilutes the sweat in the sweat gland and this change triggers the nearby nerves. What happens next has been carefully examined. The automatic part of our nervous system is very good at controlling how much blood flows to various parts of the body. In the case of wrinkling, the nerves tell the blood vessels to shrink, lowering blood flow especially to strange little balls of blood vessels in your toes and fingertips called glomus bodies. These are normally used to help reduce heat loss and keep your hands and toes warm. When the glomus blood vessels constrict, the whole glomus gets smaller, and the flesh under your skin shrinks a little. The skin on top, the *stratum corneum*, stays the same size, but since the flesh has shrunk underneath, the skin has to pucker up to accommodate this underlying change. So, it has nothing to do with water soaking into the skin, just that the flesh under your skin gets smaller and the skin on top stays the same.

As with any good bit of science, this turns out to be a rather useful observation. It was Seamus O'Riain who first suggested

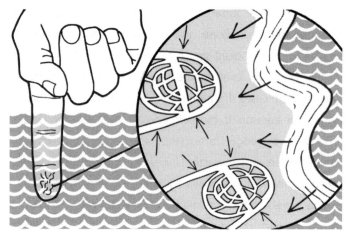

When glomus blood vessels constrict,
your fingers prune

using finger-wrinkling as a test of the health of the nervous system. There is now a standard protocol for soaking a patient's hand and checking how finger wrinkles develop over time. It is not a universally used test as it has a few problems. Firstly, when is a wrinkled finger wrinkled? How do you measure finger-wrinkling in an objective way? In addition, smoking and several routine drugs can inhibit your ability to wrinkle. These issues aside, being able to test a patient's nerve health by just popping their hand in a bowl of water is jolly handy, if you will pardon the pun.

So, that settles the big why do your fingers wrinkle debate, but it leaves one even more interesting question. Why does our body go to this effort on our behalf? Why would we have developed this strange ability to make wrinkles on our fingertips? We don't really know the answer to this yet; the only theory at the moment is that it gives us better grip.

The idea first surfaced when evolutionary biologist Mark Changizi, based in Boise, Idaho, noticed that wrinkles on wrinkled fingers looked similar to the patterns of water flow in river deltas, or even to the tread pattern on modern car tyres. In both cases this pattern, whether natural or designed, is there for a reason: to most effectively move water away from an area to the edge. To test this, another group headed by Tom Smulders at Newcastle University, in northern England, asked a group of volunteers to move a pile of marbles and fishing weights from a water-filled bowl to a box. They had to pick up each object with their right thumb and forefinger, pass it through a hole in a small screen to their left thumb and forefinger, and then place it in a box. Each volunteer had to do this both with and without wrinkled fingers. So, quite a fiddly task and one that I imagine the poor volunteers found deeply pointless, although they did get paid for their time. The results showed pretty clearly that if you had wrinkled fingers you could move the objects quicker than if you had smooth fingers. Wrinkled fingers do appear to give you more grip in the wet.

But this does not answer my question of why our bodies should go to all this effort. This is where we can dive into the woolly depths of informed guesswork. Possibly, our ancient, pre-human primate ancestors might have evolved this trait to help them cope better in wet conditions. Imagine the daily torrential downpour in a rainforest: the branches of trees become suddenly slippery and slick with wet moss and lichens, our primate ancestors turn on their special wet-weather grip

145

and hang on more effectively. Finger-wrinkling seems to be a quirk of our deep evolutionary past made present for us whenever we take a bath. The real explanation of how and why may not be the simple, elegant theory that you read so often, but it is a far more exciting and interesting one. That said, a 2013 study failed to find an improvement in grip in the wet, so the case is far from closed.

How cold are your toes?

Here's a conundrum: the normal core temperature of the human body is on average 37 °C (98.6 °F), and yet there are many people in the world, and you know who you are, that have icy cold feet and hands. How is this possible, given that the core temperature of a human varies only a tiny bit?

Over the course of a day, and even between different days, your core temperature will only change up or down by 0.5 °C (0.9 °F). Even between different people the average core temperature only varies by about 0.7 °C (1.3 °F). This, however, is your core temperature; the temperature of your fingers and toes can be significantly different. As an example, I just took a few readings on myself. I currently have a slightly low core temperature of 36.6 °C (97.9 °F), possibly because it's first thing in the morning, or maybe because I used a thermometer under my tongue, which often underestimates core temperature. What's more interesting is that my fingertips are at only 30 °C (86 °F) and my toes are a chilly 24 °C (75 °F). My hands and fingers are perfectly comfortable, but I will admit that my toes feel a bit cold right now. Perhaps I should go and dig out some woolly socks. While all of this is informative, I don't suffer from cold hands and feet. For those that do, the skin temperature can drop to below 20 °C (68 °F) – trust me, you really notice this in others if you are a warm-toed kind of person.

Heat is generated within your body by the various chemical processes taking place that release heat as a waste product. Your blood circulation then carries this heat around your body, maintaining your various bits at whatever temperature is needed, and this is the crux of the matter – what temperature do your hands and feet need to be? Since they don't contain any delicate organs and are essentially just muscle and bones, they work perfectly well at internal temperatures as low as 15 °C (59 °F), without any long-term damage. Also, since they are your extremities, they are more prone to losing heat to the atmosphere than other parts of your body, such as your torso, for example.

The way your body keeps a check on core temperature is two fold. Not only do you automatically monitor the core temperature itself, but you also unconsciously measure the temperature at your skin. If the skin temperature drops, the body makes the sensible assumption that you are somewhere cold. To protect your core temperature, it shuts down the flow of blood to your extremities. All of which stops your core temperature from dropping, but at the expense of colder extremities.

There are ways that this can become more pronounced and extreme. The first is the amount of body fat a person has. If you are carrying lots of body fat, this will insulate your body and keep your core temperature from dropping. Perversely, since your core temperature stays high even in a cold environment, your body does not push any more blood to your extremities and consequently they get even colder. Since women have on average 7 per cent more body fat than men, this is one of the key reasons that women generally have a slightly higher core temperature, but colder extremities than men.

Another factor that affects only women is the presence of the hormone oestrogen. This changes the pattern of blood flow within the body, and in particular it makes it more responsive to changes in ambient temperature. Women in the middle of their menstrual cycle often find that their hands and feet are colder than normal, while their core temperature is slightly higher than usual.

It's quite common for temperature regulation in hands and feet to go a little haywire at times. It's estimated that this happens in up to 10 per cent of the population in the United Kingdom, especially in women, probably because of the oestrogen. When it does go haywire, the most common cause is Raynaud's phenomenon, named after a French bloke who discovered it in the middle of the nineteenth century. We don't currently know why it happens, but for sufferers, any exposure to sudden drops in temperature causes the blood vessels in their hands and feet to spasm and constrict. This virtually shuts down blood flow and the affected extremity goes first white and then bluish. Even an action as simple as taking something out of a freezer can set it off. It can be really painful, especially when the blood returns. What is peculiar is that it can happen in just a part of your hand or foot, such as a single finger or toe, and for some people it only happens on one side of the body. If you have Raynaud's you need to be careful, as it can lead to long-term damage to the affected body part.

While there are medical treatments for severe cases of Raynaud's phenomenon, there is a much simpler, if somewhat prosaic, preventative treatment for those that just have cold hands and feet. The UK Raynaud's Association recommends the purchase and then wearing of hats, gloves and fuzzy socks whenever symptoms are likely to occur. This includes night-time occurrences and the wearing of the aforementioned hat, gloves and fuzzy socks in bed. Failing this, find somebody who is warm, upon whom you can reheat your cold extremities.

To dream, perchance to remember it

It's been estimated that, on average, every single one of us will spend a total of six years of our life dreaming or, to break out the maths, two hours a night over a seventy-year lifetime. This probably doesn't fit with your own experience, though, as most people do not remember dreaming every night. In fact, in studies of people's dream habits, on average, we recall dreaming only once every other night. There's that word 'average' again: in this case there is a wide spread, from those who recall multiple dreams every night to people, such as myself, who very rarely remember dreaming. However, not recalling dreams in the morning does not mean we have not been dreaming.

The science of dreams really begins in 1953, when Nathaniel Kleitman and his student Eugene Aserinsky, working at the University of Chicago, noticed that there are two distinct types of sleep. The first they dubbed rapid eye movement sleep, or REM sleep, that occurs in roughly half-hour blocks, and you normally get about four periods of REM sleep each night. Between the REM sleep periods is the imaginatively named non-rapid eye movement sleep. What makes the REM sleep stand out, and is why Kleitman and Aserinsky noticed it, is that your eyeballs move about as if you are looking at something, even though your eyelids are closed. If you wake somebody up during a period of REM sleep they will almost always tell you they were in the middle of a dream. It's now thought that the muscles controlling your eye movement respond to the visual imagery conjured by your brain as you try to look at things in the dreamscape.

We all dream, but why do some of us remember our dreams more than others?

The first and simplest reason is the natural analogue to what occurs in sleep laboratories, in which they wake people up to ask whether they have been dreaming. Anything that causes you to wake up in the night will increase the chance of you remembering your dreams. Since the non-rapid eye movement part of your sleep is the deepest, and REM sleep the shallowest, if you're going to wake up during the night, chances are it will be during REM sleep and a dream. Drinking too much liquid before you retire to bed is a sure-fire way to cause night-time trips to the toilet. The same effect, with a different cause, explains why heavily pregnant women report a huge increase in recalled dreams. Equally, eating very rich or spicy food can cause indigestion that wakes you up. So, yes, this does mean that eating too much cheese gives you dreams. Or to be more precise, eating lots of very fatty dairy products just before bed can lead to digestive issues that wake you up from periods of REM sleep, and you then remember a dream. It's worth noting that neither caffeine nor alcohol have a direct effect on dream recall. Caffeine just keeps you awake, and while alcohol can cause disturbed nights and more remembered dreams, this is usually just down to having a full bladder.

This still does not explain why, if we dream every night, we don't recall dreams every morning. Sleep researchers and psychologists have scrutinized possible correlations between different personality types and our ability to recall dreams. To quantify what makes us think and how we behave, psychologists often call upon what are known as the Big Five personality traits: openness to experience, conscientiousness, extraversion, agreeableness and neuroticism. The theory is that your personality can be rated, on a sliding scale, from a lot to a very little of each of these Big Five traits. It turns out that there is a correlation between ability to recall dreams and your personality, with people who rate highly on the openness to experience scale being much more likely to recall their dreams. All other personality-linked factors, and your sex, don't seem to make any difference.

Someone who shows openness to experience will be inventive and curious, as opposed to cautious and consistent. They might have a rich vocabulary, a vivid imagination, or be full of ideas. If you are open to experience you are interested in the world around you, appreciate art and are willing to try new things. It also looks like there is a good chance that you will be better at recalling your dreams than people who do not rate as open to experience. All of which is great, but it still does not tell us why some people recall dreams better than others. Correlation is not, after all, causation.

The only real explanation comes from a neuroscientific theory called salience. This is our ability to pick out important or salient features from the mass of information pouring into our brains from our senses. This is a vitally important trait that allows us to navigate the world. Imagine I show you a picture of somebody you know, a friend or a relative maybe. You can immediately recognize the person because your brain picks out the salient features such as hair colour, shape of face, size of nose and so on. Without this ability you would be overwhelmed by all the extraneous data about the colour of the sky, the clothing, where the photo was taken and a wealth of other information. This ability is hard-wired into our brains, but like all human characteristics, some of us are better at it than others. In extreme cases, where people assign salience incorrectly to things that are not important, you may see pathological behaviour such as schizophrenia.

So, the salience theory of dream recollection goes like this. If you are very good at picking out interesting and salient things in your daily life you are more likely to be open to experiences. When you then fall asleep, your brain conjures similarly interesting things into your dreams, which you then notice, because you are good at picking out salient things.

It should be noted that, as with so many psychological studies, the test groups used in dream science are invariably college or university students. Not only that, but most of these students are studying psychology at the same time and are paid to participate in the currency of course credits for their degree. We can only assume that what little we do know applies to humans across the globe. Chances are that if you recall your dreams it's because you are open to experiences and good at noticing interesting or salient things. This is clearly not the full explanation, though, as there are some, myself included, who do not recall dreams and yet still score highly on the openness

153

to experience scale. Which just goes to prove, somebody's always got to be awkward.

The stink of sweat

Those of you that have ever frequented a gym of any sort will almost certainly have experienced the piquant aroma of the changing room. From this you might assume that the human body is destined to be smelly after exercise. However, not all people are, and not all sweat is, stinky.

The vast majority of sweat produced by a person is 99 per cent water mixed with a bit of sodium chloride, or table salt, and a teeny amount of other trace minerals such as potassium, calcium and magnesium. When this is produced it coats your skin and then evaporates, cooling you down and leaving the salt behind. In extreme, very sweaty cases, this salt may leave a white mark on your clothes. It does not, however, give rise to any smell. So, where does the sweaty pong come from? It turns out that there are two types of sweat: there is regular, mostly water and a bit of salt, sweat, and then there is stinky sweat.

Within your skin there are millions of sweat glands. They are most densely packed on the palms of your hands, with around 350 sweat glands for every square centimetre (2,300 per square inch). Even the fronts of your knees have sweat glands, although they are more spread out at only 50 to a square centimetre (320 per square inch). All of the glands on your hands, legs, back and over most of your body are called eccrine sweat glands. These consist of a single tube rolled up into a squiggly ball that sits just beneath the surface of the skin. The tube extends from this ball, up to the skin's surface. As your core body temperature rises, your unconscious brain realizes it needs to do something. Nerves are stimulated and the sweat glands are activated into secreting water and salt. This flows up the tube, onto your skin, which is then cooled down. Blood flowing near the skin is subsequently cooled and can then flow back to the core of your body and return it to a normal temperature.

Another type of sweat gland exists though, called the apocrine sweat gland. These specialized glands are found in just a few places; predominantly your armpits and groin, but

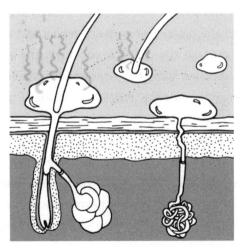

Apocrine stinky sweat gland (left) and
eccrine non-stinky gland (right)

also in other bizarre locations such as around your nipples, in the beard area on men, in your ears, at the base of your eyelashes, and up your nose. You may have spotted that the common thread to all these locations is the presence of short and curly hairs. The apocrine sweat gland does not have the elegant squiggles of the eccrine gland, instead it is just a blunt tube that feeds into the root of short, curly hairs. It is the sweat from these glands that can produce an unpleasant pong.

What makes the glands different is how they secrete the sweat. In eccrine glands, little balloons of water mixed with salt are produced inside the cells lining the tube. These balloons are shunted to the surface of the cell where they empty their contents into the tube, and this liquid is your sweat. Apocrine glands work in a completely different way, although the cells lining these glands also fill up with teeny tiny balloons of water and salt. However, rather than carefully secreting this liquid, the cells just explode, spewing their contents into the tube leading to the hair root. The sweat produced by apocrine glands contains not just water and salt but also the contents of all the exploding cells, including fat, protein and sugar. Apocrine sweat starts out as a cloudy and slightly viscous, but odourless, liquid.

Unfortunately, apocrine sweat is a feast for the bacteria that live on your skin. They immediately start to digest all the fat, protein and sugar and produce a slew of smelly chemicals, three of which give rise to the characteristic smell of body odour. Butyric acid and isovaleric acid, made from the breakdown of fats, both give rise to strong cheesy smells, and in the case of butyric acid, this is also the smell of vomit. In some cases, you also produce propionic acid, which gives a vinegary smell. Put together, these by-products of bacterial digestion are what make sweat stink.

Interestingly, there are a couple of genetic dimensions to smelly sweat. There is a single change in a single gene, with the confusing name of ABCC11, that is common in East Asians. This genetic difference has two effects. Firstly, it makes earwax dry and flaky rather than sticky, which while interesting appears to have no physiological impact. However, people with this genetic difference also have far fewer apocrine sweat glands and, consequently, have much less stinky sweat.

On top of this, it looks as though the precise chemical make-up of your sweat may influence your choice of partners. As part of our immune system we each have a genetically defined, unique set of molecules on the surface of our cells called the major histocompatibility complex, or MHC for short. It would appear that when choosing a partner we literally sniff out people who have an MHC that does not match our own. In theory, this would mean that offspring with such a partner would have a more varied MHC and possibly be healthier. This would partly explain why we have evolved to have not one, but two types of sweat.

However, none of this explains the source of the stinkiest sweat smell. I am of course talking about gym shoes, plimsolls, sneakers, trainers or whatever terminology you prefer. These are usually quite repulsively smelly if worn regularly, particularly without socks, and yet your feet have no apocrine sweat glands.

Surely, since you only have the water-and-salt-producing eccrine glands on your feet, your shoes should not smell. Unfortunately, while this is true, when you walk, or vigorously exercise, your feet also shed large amounts of skin cells. Add this detritus to a damp, enclosed trainer, and bacteria are once again provided with nourishment and the opportunity to produce the characteristic, and horribly odiferous, by-products of bacterial digestion.

How to grow a new limb

I was trimming my fingernails recently and it set me wondering. Why was it that I could cut my nails short and they would regrow, while my fingers were an altogether different matter? If I chop off even a small part of my finger it will not grow back at all. It will heal over with scar tissue and that's the end of it. In fact, aside from my liver, I can't regrow any of my own major body parts. However, there are animals in the world that can do this and will regrow body parts that have been lost. Why is it that they can regrow limbs, but I can't? It would be a rather handy trait to possess (pun intended).

The regeneration of body parts has been a popular theme in the history of science fiction stories and comic books, yet it's an even older story in the annals of science. In 1744, a Swiss tutor called Abraham Trembley, while teaching the children of a posh Dutch politician, noticed some peculiar microscopic creatures in a sample of pond water. He named these tiny blobs of jelly *hydra*, because the numerous tentacles sprouting from one end reminded him of the many-headed monster of Greek legend. Not content with just observing these strange new creatures, he decided to experiment on them. In the tradition of eighteenth-century science, the obvious thing to do was to chop one in half and see what happened. To his surprise, and the shock of the scientific community when he reported his findings, the hydra did not die, but instead recovered and quickly grew into two new hydra. Each half had regenerated into a whole.

Since then a number of different types of animal have shown the ability to regrow limbs or to regenerate damage. The best-known example of this is probably that of a type of lizard called a gecko, that has a couple of remarkable abilities. Firstly, it can walk up vertical walls, which is very cool but nothing to do with regeneration. More relevantly, it can also shed its tail when attacked by a predator, and then grow a new one over the course of a few weeks. In fact, geckos are really good at regrowing bits of themselves. They've been known to regrow legs, jaws, internal body parts and even their eyes. It would seem that so long as the injury does not outright kill a gecko, it can regrow the bits that were lost or damaged.

We know a bit more about regeneration from work using salamanders that are closely related to geckos. Salamanders share the regenerative abilities of geckos, although they can't climb a vertical wall, so they are intrinsically less cool. If a salamander loses a leg, a layer of cells form at the surface of the lost limb that then signal to the underlying tissue. The cells in this underlying tissue then revert to the sort of cells found in the salamander embryo. These are what are called stem cells and they have the potential to grow into all sorts of different types of tissue, such as muscle, bone, nerves and skin. Stem cells that are capable of making lots of different types of tissue are really unusual in an adult animal. Within a human, there are a few types of cells that can do this to a limited degree. Bone marrow cells, for example, can turn into any of the dozen or so types of blood cell, both red and white, but within an adult human body, there are no cells that can create the range of tissues needed to regrow a limb. How the skin over the stump of the salamander's lost limb transforms the cells underneath is, as yet, a mystery. The crucial signal has not yet been identified.

On top of that, we also don't understand how the regrowth of the limb is co-ordinated. Once the stem cells have been made, we don't know what makes some turn into muscle and some into bone. Furthermore, it's a mystery how the cells know where they are in the limb and when to stop making leg and start making foot. There remain a huge number of unanswered questions.

No examples of this were known in mammals, until that is the spiny mouse came onto the scientific scene. In 2012, developmental biologists at the University of Florida published their work on a couple of species of African spiny mice. These rather cute little mice have stiff hairs in their coat, hence the spiny name, but also, rather gruesomely, can shed great lumps of skin right down to the muscle underneath in response to predators. The mice then regrow the skin without any visible scarring.

While it is possible for some mammals, lizards and blobs of pond jelly to regenerate lost body parts, we humans can't do it, at least not yet. It seems to be something that the human body is particularly set against. One theory is that regeneration becomes a fatal trait when you have a much longer life span when compared to a tiny mouse. Having a system that allows your adult cells to start growing again could lead to potentially lethal cancers if it went wrong. Since the longer you live the higher is the risk of this happening, our evolution has shut down any possibility of salamander-like regeneration.

The dream of being able to regenerate our bodies is still a long way from reality. Our evolution as long-lived mammalian vertebrates means that our ability to create stem cells has been locked down. Consequently, we no longer have the ability to regrow many of the different types of tissue that make up our bodies and regrowing a limb is not currently possible for us. Unless, of course, we can develop the serum used by Curt Connors in the SpiderMan universe. Unfortunately, while this did allow him to regenerate his arm, it also turned him into an evil super-villain lizard. I suspect that the drug regulatory authorities are probably not going to be too happy about this as a side effect.

Science in the World Around Us

Platinum-paved roads

Lurking underneath your car, usually just beneath the engine, is a steel box hooked up to the exhaust system. These boxes first began appearing on cars back in the mid-seventies but now they are ubiquitous, with every car, van and lorry carrying one. Inside each box is a ceramic honeycomb structure coated with about 3 or 4 g (about ⅛ oz) of platinum and the equally shiny and expensive metals called palladium and rhodium. These metals help chemical reactions get started within the exhaust fumes, converting potentially harmful gases into predominantly harmless ones. The platinum and other metals act as catalysts: they aren't consumed by the work they do and you need very little of them to give the chemistry a kick-start. It's the presence of these precious metallic catalysts that explains why the steel box under your car is known as a catalytic converter. It's also why a new catalytic converter can set you back a large amount of money.

Disappointingly, especially given the promise of the platinum, the inside of a catalytic converter is a murky grey or brown. Most of them contain what is known in the trade as a cordierite monolith, a single block of extruded ceramic about 20 cm (8 in) long and maybe 15 cm (6 in) in diameter. The block is not, however, a solid lump. It's shot through with thousands of tiny tubes, usually square in cross section, only 1 mm (about $\frac{1}{32}$ in) across and arranged in a beautiful, regular pattern. In some cases the monolith is made of corrugated metal sheet, but the end result is the same: an inert block, with a large surface area through which exhaust fumes can flow.

The platinum, palladium and rhodium come into this story when the monolith is drenched in a solution containing these precious metals. They coat the inside surfaces of the tubes and, when dry, leave behind a microscopically bumpy surface. The point of all this effort is to create a maximally large surface area coated in platinum, palladium and rhodium, and the key to the reason why is down to how a solid can help along, or catalyze, a reaction that takes place in gases.

Exhaust fumes from petrol engines contain a number of toxic gases. The most well known is carbon monoxide, a poison and a greenhouse gas, but there is also unburnt fuel that acts as a major air pollutant. Possibly most unpleasant, though, are the oxides of nitrogen that give rise to acid rain and destroy the ozone in the Earth's atmosphere that helps to filter harmful ultraviolet radiation from the Sun. Modern catalytic converters deal with all three of these. The platinum and palladium help oxygen react with carbon monoxide and the unburnt fuel to produce harmless carbon dioxide and water vapour. Meanwhile, the rhodium and the platinum catalyze the breakdown of the oxides of nitrogen to make nitrogen gas and oxygen. For these reactions to happen the gases need to be in physical contact with the precious metals, which is why you need the ceramic monolith with all its skinny air tubes. If you just had a block of platinum, most of the gas would flow over the block without touching it or reacting.

There are a few minor problems with modern catalytic converters. For the first reaction to take place, with the carbon monoxide and unburnt fuel, you need oxygen for it to work and it is quite possible to tune a car so that there is very little oxygen in the exhaust fumes. Modern cars monitor the oxygen levels, before and after the exhaust enters the catalytic converter, and adjust the air to fuel mixture injected into the engine accordingly. Another issue is that all of this chemistry only happens at high temperatures, usually well above 400 °C (750 °F), and it takes about five minutes for the converter to reach this temperature. So, the pollution created by short journeys does not get converted into less harmful components, because the catalytic converter does not work until it has time to heat up.

Catalytic converters are also susceptible to being poisoned and permanently ruined by such issues as leaks from coolant systems and lead in the fuel, which is why they only started appearing once unleaded fuel became common. The biggest problem, though, is that catalytic converters in cars eventually wear out. This is not because the precious metals get used up doing their job – remember that catalysts are not consumed by the reactions they start. Instead, driving causes vibrations and

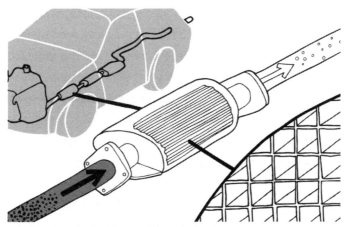

Inside a car's catalytic converter

knocks that shake free the coating of catalysts on the ceramic monolith of the converter. The reason that you ultimately need to fork out for a new catalytic converter is because the precious metal coatings have been shaken free from your old converter, lost from the end of the exhaust pipe, and are now strewn all over the road.

Which means that modern road dust is chock-full of platinum, palladium and rhodium. If you take the sweepings from urban roads and pick out all of the gross bits – the plastic wrappers, tin cans and organic material – what you are left with is a dark brown sludge. Much of this has come from our cars wearing out. Anyone who has driven a car knows how often expensive tyres need to be replaced. Where do they go? The ground-up rubber ends up on the roads along with those metallic particles from catalytic converters.

The best platinum mines in the world dig out ore with just

a few parts per million of the precious metal. It's expensive, dirty, environmentally damaging work, but it is worth the effort of mining as the platinum that is extracted is so very, very valuable. The brown sludge swept from the roads has the same sort of levels of platinum as this best mined ore. The process for purifying the platinum, palladium and rhodium from road sweepings is only just being developed by scientists at the University of Birmingham in the United Kingdom. In the UK alone, there could be tens of millions of pounds' worth of precious metals lying on the streets, if only we could get at it.

The slippery ice conundrum

Water in all its forms is a most peculiar thing that defies the behaviour we expect from well-behaved substances (see page 68). What makes this all the more peculiar is that it's also, arguably, the chemical substance with which we are most familiar. Water quite literally permeates our everyday lives, yet it retains a few mysteries. Foremost of which is why frozen water, or ice, is slippery.

There is a classic explanation for the slipperiness of ice that used to be very common in textbooks and on the Internet, and it goes like this: when you stand on ice you apply pressure that causes it to melt, creating a layer of lubricating liquid water that you slip on. Traditionally, this explanation is accompanied by a demonstration that shows how pressure does indeed cause ice to melt. A block of ice, supported at both ends has a wire looped over the block. Each end of this wire is then attached to heavy weights. These weights create high pressure under the wire, where it lies on the ice. The high pressure lowers the melting point of the ice just beneath the wire to below freezing point and the ambient temperature of the block. As a result, the ice under the wire melts and it very slowly cuts through the block of ice. If it's done right, as the wire cuts down into the block the water created by the pressure refreezes above the wire and the block remains whole. The wire passes through the ice block, leaving it intact.

Unfortunately, if you actually do the mathematics of a real-life slippery example, say a person stood on an ice skate, it does not add up. The pressure created by an ice skate is only a tiny fraction of that created by the artificial situation of the weighted wire. In the case of an ice skate, the melting point of the ice is only reduced by a minuscule amount, to about −0.02 °C (31.97 °F). Since skates remain slippery on ice at temperatures well below zero, the pressure theory cannot be what makes the ice slippery.

In the last few decades, a couple of different explanations have come to light, both with experimental data to back them up. In 1996, Gabor Somorjai at the Berkeley Lab in California decided to tackle the question of what makes ice slippery. He started with the reasonable assumption that ice is slippery because the surface is lubricated by liquid water. Our everyday experiences of ice back this up, as immediately you pull an ice cube out of the freezer it has a wet and slippery surface. However, examine an ice cube in temperatures well below freezing and there is no visible sheen of water. The ice looks rock solid even on the surface, but it's still slippery. Dr Somorjai decided to test this by firing beams of electrons at very cold ice. In theory the surface should have been 100 per cent solid, but the reflected beam of electrons produced a pattern associated with a liquid surface. The explanation that emerged is that the molecules of water at the surface of the ice are just not held tightly enough, and they are free to move about as they would in a liquid. This additional movement creates a layer of liquid water a few molecules thick on the surface of even the coldest ice. While this layer would be too thin to see, it should be enough to make any ice slippery no matter what temperature.

So far so good, but then a few years later, in 2002, Miquel Salmeron, a colleague of Dr Somorjai's at the Berkeley Lab, used a funky piece of kit called an atomic force microscope to measure the friction of the surface of ice. This particular type of microscope uses something a bit like a tiny record player needle that is dragged over a surface to make an assortment of measurements. In particular it can give you an idea of how rough a surface is on a very microscopic scale. What Dr Salmeron found was that the surface of ice on this scale was not smooth, but very rough. In which case, the slightest movement on the ice creates friction with the rough ice surface, and all friction results in the generation of heat. This, Salmeron suggested, is what lubricates the ice. The microscopic friction creates heat that melts the surface of the ice, making it slippery.

In the end we have one outdated theory and two possible explanations, both backed up by hard data. While neither of these two ideas contradicts each other, equally neither supports the other. So, why, then, is ice slippery? The jury is still out on this question. What I find most bizarre is that such a very ordinary question still has no definitive answer. Maybe it's a mixture of the two theories, or maybe there is something else going on altogether. Whatever the case, next winter when you slip on the ice and land painfully on your rump, you can delight in knowing that, while science acknowledges what happened, it can't explain exactly why.

Converting your electricity

In the room in which I currently sit there are a total of eight electrically powered devices. Of these, six are plugged in through power convertors that change the alternating current (AC) of the mains to a direct current (DC) of 12 volts. The seventh device, my computer, also runs off direct current, although the power pack is built into the body of the device itself. The last item is a paper shredder sat at my feet and, while I'm not absolutely certain without dismantling it, this does not have a power convertor. It's the only device in the room that runs off alternating current. This pattern is repeated around my house; most electrical devices run on direct current, rather than the alternating current that is supplied to all the household sockets.

This may strike you as a rather odd situation. The power convertors we all have in our homes are at best 90 per cent efficient. In a 2001 speech, President George W. Bush called power packs *energy vampires*, as not only do they waste 10 per cent of the energy going into them, but the older varieties used power just by being plugged in, even when they weren't actually powering a device. The wasted energy is lost as heat, which is immediately noticeable when you go to unplug them. It would seem to make sense to have a direct current supply of electricity in our homes.

This is not a new idea; the supply of domestic direct current predates that of alternating current. Back in the 1880s, two giants of American industry battled it out over how we would supply electricity to our homes. Thomas Alva Edison, the great inventor and entrepreneur, stood on one side of the War of Currents. He advocated direct current, claiming that it was safer and more useful. The direct current electric motor, for example, had long been developed into an efficient and practical machine; conversely, the alternating current motor had only just come off a drawing board. Edison went to extraordinary lengths to discredit his rival, George Westinghouse. He embarked on a no-holds-barred smear campaign against alternating current, publicizing accidents and filming the alternating current electrocution of stray cats and dogs. He even went so far as to invent and build the alternating current electric chair that was used

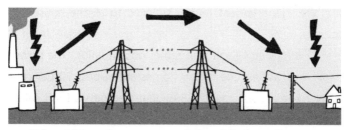

Using AC transformers to send electricity long distances

for the execution of condemned prisoners. In an even more unpleasant twist, he dubbed the use of his electric chair as 'being Westinghoused'. However, George Westinghouse, collaborating with figures such as Nikola Tesla, won in the end: as nobody wants to live next door to a power station, you need to be able to transmit electricity over long distances.

Whenever you let electricity flow along a wire, no matter how thick, or how conductive, there will be a loss of energy. This loss of energy depends on the current flowing in the wire. Double the current and the energy lost goes up by a factor of four, halve it and the loss goes down fourfold. Also, for a given amount of power, as the current goes down, the voltage goes up.

Taken together, these two fundamental facts of electrical physics impact how we send electricity long distances down wires. To minimize energy losses we need to use a low current, but to deliver a decent amount of power at this low current, the voltages need to be very high. The system we currently use for the transmission of electricity from power stations, along overhead power lines, has voltages that exceed 765,000 volts. This makes for efficient transmission of power, with minimal energy losses. The difficulty is the creation of such high voltages.

With alternating current, it's dead easy to convert low voltage to high voltage and back again. Michael Faraday invented the transformer that does this some fifty years before the War of Currents. By using a transformer, we can multiply the voltage at the power station to hundreds of thousands of volts, send the electricity to a small, local sub-station, then bring it back down to a sensible voltage and feed it to your home.

If you tried to send 12 volts direct current from the power station to a house just 1 km (about ½ mile) away, you would need to reduce the resistance in the cables to such an extent that the wire would be an impractical 50 cm (20 in) in diameter. Even for a moderately sized house with a source of 12 volts of direct current in a garage, you would need wires four times as thick as those you currently possess. It turns out that it is much easier to use alternating current right up until the point where you need the direct current.

I'm afraid that if you are not a fan of your collection of assorted power converters, there is nothing you can do. Unfortunately, all our electronic gadgets and gizmos rely on delicate silicon microcircuits and chips which, because they only work with electricity moving in one direction, must use direct current. Take heart in the knowledge that not only are modern power convertors more efficient, smaller and don't use any energy when not in use, but also that their inevitability is determined by the fundamental laws of physics.

Car-seat electrification

The typical spark made when you clamber from the car is about 1 cm (about ½ in) in length and in dry air that can easily be 20,000 to 30,000 volts of static electricity. Twenty thousand volts sounds like a very large and dangerous jolt of electricity, and yet we have all experienced this and probably much greater voltages without suffering any long-term negative effects.

Our understanding of static electricity reaches way back into the history of science, back to at least 600 BCE, when it was first mentioned by the Ancient Greek philosopher Thales of Miletus. He noted that when amber was rubbed against such things as cats, it produced tiny crackles and sparks. He pondered on the meaning of this and presumably upset a number of cats along the way.

We had to wait over two thousand years until the late nineteenth century for a real understanding of why cats and amber created sparks. At the core of all electricity is the electron that was finally discovered and identified by Joseph Jon Thomson, a British professor at Cambridge University. He realized that what made electricity flow, and sparks jump, was a build up of these unbelievably minuscule, sub-atomic electrons that each carry an electrical charge.

When Thales was rubbing his cat with a lump of amber, electrons from the cat's fur were being transferred to the amber, making the cat slightly positively charged and building up a negative charge on the amber. Eventually, when the difference in charge was great enough, sparks would fly between the two

173

and the cat would presumably cease cooperation. It turns out that it is not just cats and amber; many substances are good at giving up electrons and many are happy to accept them. Scientists of an electrical persuasion have long played around with this and created the triboelectric series, where the *tribo* part comes from the Greek word for 'rub'. This series lists a whole range of materials, from those that are very good at giving up electrons, and tend to become positively charged when rubbed, down to those that are very good at accepting electrons, and become negatively charged. Almost at the top of the list for giving up electrons is human hair, a bit further down are cats, and much lower down the list, well into the electron-receiving section, is the sort of rubber from which balloons are made. Which is why if you rub your hair with an inflated balloon you become charged up with static electricity and your hair stands on end. The hair has become positively charged as electrons have been rubbed onto the balloon. Since positive charges repel, strands of your hair are pushed away from each other and they stand on end. They are also attracted to the negative charge on the balloon.

We have all experienced this build-up of electrical charge, and we know it as static electricity since it is just that – static and unmoving. Electricity from a battery or from the mains socket flows along conductive metal wires, but static electricity forms on non-conductive things such as hair, rubber and cats. We only know it's there when the hairs start to stand on end, or we get zapped by a spark. It turns out that air is not a completely useless conductor of electricity. If enough static charge builds up between two nearby objects, it will eventually reach across the gap, neutralizing the positives with the negatives. When this happens, the friction caused by the flow of electrons heats up the air to an amazing degree and creates a white-hot stream of super heated molecules. The sudden heating and then cooling of the air makes the cracking noise you hear. Depending on how humid the air is, for every centimetre of spark you need a voltage of between 15,000 and 30,000 volts (about 50,000 volts per inch).

So, how does all this Ancient Greek cats with amber stuff relate to the spark you sometimes feel when you step from a car? Well, the seat in your car is designed to be hardwearing and comfortable to sit on; to that end, car-seat designers usually choose to cover their seats in polyester cloth, or sometimes a vinyl-coated fabric. You, on the other hand, are probably wearing more comfortable materials, such as cotton, wool or nylon. If you check your handy triboelectric series, both polyester and vinyl are very low on the list and tend to become negatively charged. Your comfy clothes, however, are higher up the list and will readily give up electrons and become positively charged. As you go to get out of your car, you swivel in your seat, swinging your legs out and rubbing your cotton denim, jeans-covered posterior against the polyester seat material. This action alone creates a huge transfer of electrons from you to the car. Since the only part of you touching the car is your hand holding the plastic, non-conductive door handle, that charge stays on you as you step out of the car and lift your bottom from the seat. You are now standing on the ground, probably wearing insulating, rubber-soled shoes, and, again, the positive charge you have built up stays on your body, especially if it is a dry day. Until that is, you go to push the door closed. As your fingertip reaches for the metal of the door, electrons leap from the car to neutralize your positive charge. The spark that this creates can easily be 1 cm (about ½ in) long and thus up to 30,000 volts on a dry day. Unfortunately for you, your fingertips are crammed with nerve-cell endings that the spark stimulates, and you feel a short, sharp pain.

This might only be a momentary and trivial pain but it can lead to much more serious consequences if it happens when you are filling your car with fuel. An empty fuel tank is full to the brim with fuel vapour, and when you refill the tank all that vapour is pushed out. In some countries, such as the USA, once you have started to pump the fuel it is possible to lock the handle of the pump, allowing you to step away and let it do its job automatically. Unfortunately, people have been known to then return to the warmth of their car, and sit down to wait. When they get up, they charge themselves with electrons by rubbing their clothes against the car-seat fabric and only discharge themselves when they touch the handle of the petrol pump, creating a spark. All the displaced fuel vapour from the tank can then promptly – and terrifyingly – ignite.

There are a couple of things you can do to avoid creating a static spark when you leave your car. The first is to make sure that as you clamber out you are touching part of the metal surface of the car. The easiest way to do this is to place your hand on the metal of the car body on the side or top of the door-frame as you climb out. Inevitably, the one day that you will forget to do this will be the really dry day when you are wearing lots of cotton, and you will get a real belt of a spark from the car. However, my second, slightly impractical and possibly uncomfortable, suggestion is guaranteed to work: only ever wear clothes made from polyester and vinyl.

Keeping the greenhouse warm

We have all experienced a roasting hot conservatory or a poorly ventilated office block that swelters on a hot sunny day. Less often, but equally within our experience, is the eponymous greenhouse that suffers from this greenhouse effect. What is not obvious, I believe, is why this should be the case. Surely if the heat can get into the greenhouse, it can also get out? And yet it doesn't – the heat gradually accumulates and temperatures rocket.

At the heart of this is an interesting, if complex, bit of science called black body radiation. The physics of this is fiddly, but essentially boils down to the wavelength of the light, or electromagnetic radiation, emitted by something depending on its temperature. The average temperature of the surface of the Sun is around 5,500 °C (9,900 °F) and accordingly it emits electromagnetic radiation most intensely with a wavelength between 300 and 700 nanometres. This is a distance between

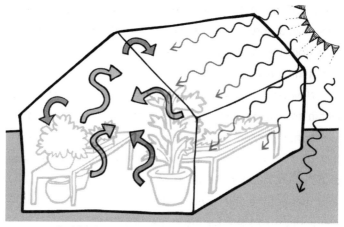

Blackbody radiation and a shift in wavelength
heats your greenhouse

the peaks of adjacent waves that is virtually impossible to grasp, but in 1 mm you can fit 2,000 of these wavelengths. It should be noted that this range of wavelength constitutes visible light, and this is no coincidence. The human eye has specifically evolved to take advantage of the part of the light spectrum that is most intense at the surface of the Earth. There is also some infrared and ultraviolet light, but much of this is reflected or absorbed by our atmosphere, although what does reach us in the case of ultraviolet light causes our skin to tan (or to burn, in my case).

When this light reaches a window, the visible light passes through, and any remaining infrared or ultraviolet is absorbed by the glass (for more on why some things are see-through, see page 85). That visible light then hits all the objects within the conservatory or greenhouse, like the wicker furniture and pot plants. Clearly, these objects are not 100 per cent reflective, so they must absorb some of the light. As they absorb the visible light from the Sun, they capture the energy of this light and turn it into heat.

Now we have to invoke black body radiation. All the objects within the greenhouse or conservatory emit electromagnetic radiation. If they are at a temperature of around 15 °C (59 °F), they give out radiation in the infrared region at a wavelength of between 7,000 and 20,000 nanometres. If you recall, the light that got the objects hot in the first place was 300 to 700 nanometres, so there has been about a twenty-five fold shift in wavelength, and this is what creates the greenhouse effect.

The air around us is much less transparent to electromagnetic radiation in the infrared region. Consequently, this infrared radiation does not whizz straight back out of the conservatory, but instead is soaked up by the air, which it warms. By shifting the wavelength of the light, it has effectively been trapped inside the conservatory or greenhouse. Once the air is being heated, you start to set up convection currents. The warm air rises, moving away from the objects heated by the Sun and allowing cooler air to move into contact. This air is warmed and the cycle continues, gradually heating up all the air. What results is a runaway heating system: the sunlight keeps adding more energy to the greenhouse, and the objects inside continue to be warmed, passing the heat to the air and the temperature climbs higher and higher.

This is essentially also the explanation for the other, more newsworthy greenhouse effect that is the current driver of climate change on our planet. The subtle, but vital, addition to the science is that carbon dioxide gas is particularly good at absorbing the re-emitted infrared radiation. Also, while we clearly don't have a glass roof on our atmosphere, the warmer air is still trapped in place by the Earth's gravitational pull. Accordingly, the carbon dioxide produced by our industrial habits is driving the temperature of the planet inexorably upwards.

For the planetary-scale version of greenhouse effect, beyond reducing carbon dioxide emissions, we don't yet have a solution. However, for domestic cases of the greenhouse effect, an understanding of the science does provide an answer. Either stop the sunlight entering the space by putting up blinds, or use the convection to create a cooling breeze. If you put a window or vent at the very top of your glassed enclosure, and allow air to enter lower down from outside, the hot air will escape and pull cool air in. It won't stop the greenhouse heating up, but it will calm down the runaway greenhouse effect.

Demisting with cold air

As winter approaches and night-time temperatures drop, the morning ritual of demisting the car windscreen begins. Next time you sit there waiting for the car heater to do its job, ponder what is going on.

Demisting is all about thermal energy. Imagine a glass of water at 20 °C (68 °F). This temperature corresponds to the average energy of all the molecules in the water. However, some molecules have more and some less energy than this average. Every now and then, a high-energy molecule reaches the surface of the water and escapes the liquid, evaporating to become a molecule of water vapour gas. Equally, water vapour molecules in the air with low levels of energy occasionally bump into the water and stay there, condensing and changing from gas to liquid. The number of water molecules in the gas above the glass of water depends on the balance between these two processes. Increase the average temperature and more molecules escape; reduce the temperature, and more condense.

Now, ditch the glass and jump back into the car, but wind back time to the night before the misted-up windscreen. When you parked, the car was toasty warm and the windows free of any mist. There's always a fair amount of water in the car, brought in on your shoes, a wet dog or simply in the air you have breathed out. Since the car was warm, the balance between evaporation and condensation favoured water turning into a gas. As you got out, the car was full of water vapour, and maybe the smell of damp dog.

Now jump forward to the next morning. Overnight, the outside temperature drops and begins to cool the car from the outside. The windows are the first things to get cold on the inside, since they are the thinnest part of the car. The air, next to the windows, starts to cool and the energy of the water vapour in the air drops. This pushes the balance towards the condensation side and water vapour molecules start to join up, forming tiny droplets of liquid water. They're encouraged to do this by specks of dust or grease on the glass windows. When you come to your car in the morning, the inside of these windows is now coated in a layer of tiny water droplets.

So, is it possible to avoid this? Well, you can't change the Laws of Thermodynamics, but there are a few things that can help. First, try to keep the inside of the car dry. Make sure your heating system is not set to recirculate, but rather to draw in fresh air that is probably less humid, and keep the windows clean, and free from greasy fingers. While these precautions should reduce the number of misty windscreens you encounter, you will inevitably find that you have misted up. At which point, when you hop in the car, you may think that the quickest way to clear the windscreen is to put the heating on full blast. However, your car heating system does not blow hot air until the engine has itself heated up, so initially it will be blowing cold, moist air that will struggle to demist anything. Your best bet is to put the air conditioning on. While it may seem counter-intuitive to blow cold air, the air-conditioning unit dries out the air as it cools it. This drier, colder air will shift the balance towards water vapour and the windscreen will gradually clear.

Boomerangs, sticks that come back

A little while ago, I had the opportunity to be part of a team attempting to break the Guinness World Record for the largest returning boomerang. The boomerang we were using was a monster at 2.94 m (about 9 ft 8 in) from tip to tip. It was made from lightweight wood, and while this meant it didn't weigh much, it was still a significant strain to throw. It was also very delicate for such a huge object, which meant that one wrong throw and it would crumple on impact with the ground. The rules for this particular record state that the boomerang must travel 20 m (65 ft 7 in) away from, and return to within 10 m (32 ft 9 in) of, the thrower. Since we had chosen the renowned and prestigious Oval cricket ground in the centre of London for the attempt we only had a very limited window of time in which to break the record. It was a nerve-wracking experience as not only was the clock ticking, but the groundsmen were keeping a close eye on us in case we made a mess of their painstakingly manicured pitch. In retrospect, while it was a spectacular place to attempt the record, it may not have been the most sensible.

Despite these issues, Adam McLaughlin, of the British Boomerang Society, and I did manage to set the world record. I will admit that, sadly, the new record was not one of my own throws. While I managed a (6.09-m) 20-ft throw, it didn't return quite far enough. In the end it was Adam who made the record throw, which was right and proper as it was he that made the boomerang in the first place. The experience opened my eyes to the world of returning sticks, and since then I have dabbled with boomerangs and explored the science that makes them return.

The first thing that I got wrong, like most people new to boomerangs, is that you don't throw them like a frisbee. Instead, you hold a boomerang vertically when you are launching it. You also have to get it the right way around. One face of the wings of a boomerang is flat, while the other side is curved into the shape of an aeroplane wing. Which is to say that one edge of each wing is fatter, while the other edge is a thinner blade. You hold a boomerang vertically, in your right hand between forefinger and thumb, with the curved surface towards you and the flat surface away from you. This technique is vital to make a returning throw and, yes, this does mean that left-handed boomerangs can be made. With a sharp

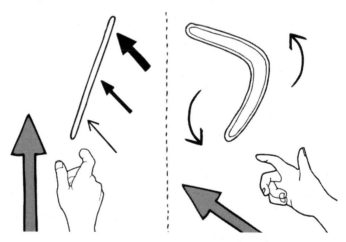

Hold the boomerang vertically with the curved surface towards you and flick your wrist to launch it

throw and a quick flick of the wrist, you launch the boomerang and give it the essential spin.

This is where things begin to get a little bit counter-intuitive, as the physics of spinning objects is a bit weird. There are two things at work with a boomerang, and the first is to do with gyroscopes. With any luck you have an idea of what I mean because you have played with toy gyroscopes at some point. These are the small wheels mounted in metal wire frames that you can spin using a bit of string and then balance on your finger. I'm hoping you are familiar with these because, if you have never seen the weird things gyroscopes can do, it seems almost magical.

Any object that spins acts like a gyroscope, and boomerangs are no exception, but let's step away from boomerangs for a moment and consider a different, more generic spinning object. Image you have a wheel that is spinning vertically around a horizontal axis – this could be a bicycle wheel or a toy gyroscope. If you now try to twist the axis down towards the ground, so that the wheel is spinning at an angle to the vertical, something very peculiar happens. The wheel will resist your attempts to twist it. The physics at play here is the law of conservation of angular momentum. When you try to twist a spinning thing there is a force that tries to stop you, but even more peculiarly, the force you are applying is rotated through ninety degrees. As you try to twist the axis of your spinning wheel down, it remains vertical and rotates to the left or right depending on the direction of the spin. This is called gyroscopic precession and really has to be seen to be believed.

Now, keep this in mind while I tell you about the second bizarre thing going on with a boomerang and its aeroplane-shaped wings. Boomerangs usually have just two wings for simplicity, but they can have any number. Three-winged Y-shaped or four-winged X-shaped boomerangs are particularly effective, and easier to throw than the traditional ones. When these are spinning in the air, the wings create lift, just as an aeroplane wing would. However, since the boomerang is spinning vertically, the lift produced by the wings pushes the boomerang not up, but to one side. If you are throwing a right-handed boomerang, the spinning wings push it to the left. The faster the wings move, the more they push the boomerang to the side.

The boomerang doesn't just spin; it also moves forward through the air. You have to add this forward speed of the boomerang to the speed of the spinning wings. When a wing is at the top of the boomerang, it spins in the same direction as the entire boomerang moves, so you add the two speeds. Conversely, at the bottom the spin is opposite to the movement, so you subtract the speeds. All of which means that the arms of a boomerang are moving faster at the top than at the bottom, and the push generated by the spinning aeroplane wings is greater at the top than at the bottom. The boomerang is not only pushed to one side, but the axis of the spinning boomerang is being twisted downwards.

The spinning boomerang acts like a gyroscope being twisted downwards, and gyroscopic precession comes into play. By this effect, the vertically spinning boomerang is pushed around in a big loop and, if you throw correctly, it comes back to the thrower.

I say correctly, as getting a boomerang to come back is a lot trickier than cartoons would have you believe. Knowing why a boomerang should come back to you probably won't help you to do this. You also need to throw the boomerang at the right angle into the wind, tweak the wings to give enough lift and perfect your wrist-flicking technique. Once you have all of that sorted out, in your hands a boomerang will cease to be a bent stick and return time and time again, thanks to the peculiarities of spinning things and gyroscopic precession. I don't, however, suggest that you use a monster boomerang 2.94 m (9 ft 8 in) across. Start with something a little more reasonable.

Garden Science: Wildlife on your Doorstep

The glue that keeps the apple crisp

I eat a lot of apples. They are probably my favourite fruit, and definitely the one I eat the most, but I also like a bit of diversity. For this reason, I enjoy the autumn and the months of winter in which it is possible to get a range of different apple varieties. One that appeared on my radar just this year is the Evelina apple. It's a variety that was only introduced a few years ago and produces large red and yellow fruit. Crucially for my tastes, the flesh of the apple is crisp, juicy and sweet.

I say crucially because I can't stand apples that are mealy, or have what is known as creamy flesh. While it is hard to describe the effect when you bite into such an apple, I'm sure you know what I'm talking about. The outside of a mealy apple may seem firm and look appetizingly red, but the flesh inside turns to a dry mush in your mouth. My sincerest apologies if you like this sort of apple; what can I say – you're wrong. Assuming I haven't offended you and you've slammed this book shut and gone off in a huff, it's worth considering why two apples can have such a different texture.

The flesh of an apple is composed of millions of plant cells that can be up to a 0.25 mm (about ¹⁄₁₀₀ in) across. There are a couple of important features of these cells that are relevant to the texture you experience when you chew an apple. Within almost all plant cells is a huge bag of watery stuff called a vacuole. The cells of an apple are almost entirely taken up by this vacuole, and the liquid contained is full of sugar. When the cells of the flesh are broken open this liquid floods out and the experience in your mouth is juicy and sweet.

There is, however, more than one way to break apart the flesh of an apple. Surrounding all plant cells is a cell wall made up of chemicals such as cellulose and lignin. It's the cell wall that gives individual cells their structural strength and integrity. Essential to the crispness of an apple is the way individual cells are glued together. A mix of chemicals form the glue, but the most important is pectin; and yes, this is the stuff that makes jam set. Pectin is a complicated arrangement of lots of sugar molecules linked together into long chains. When these chains stick to each other they form a jelly, and it's this that makes the glue that binds the cells together.

How strong and how much there is of the pectin glue ultimately determines how crisp or mealy your apple is. Strongly bound apple cells give you a crisp texture, so that when you eat the apple you need more force to split open the cells and release their juice. On the other hand, if the glue is weak, when you bite into the apple the cells break apart from one another

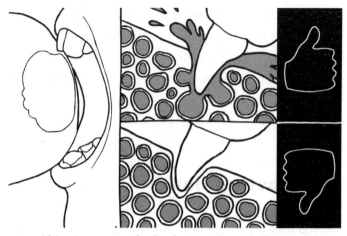

Biting into a crunchy (top) or mealy apple (bottom)

rather than breaking open. Now, instead of lots of smashed-open cells and juice, your mouth is full of clumps of unbroken cells. When the pectin is weak, the apple has a mealy texture.

So, what determines if an apple is going to be mealy or crisp and juicy? In the end this comes down to the genetics of the particular variety. All apples start out as really hard, sour balls on the tree. As they ripen, sugar is transported into the fruit, changing it from sour to sweet. At the same time, proteins or enzymes produced inside the cells of the apple are transported to the cell walls. Here, the enzymes change the pectins gluing the cell walls together, reducing the adhesion between the cells.

Different varieties of apple will take these two processes to different lengths. Some, such as Gala apples become really sweet and retain strong pectin bonds between cells so have a crisp juicy bite to them. I quite like Gala apples. Other varieties, such as McIntosh apples, don't develop such high sugar levels but the enzymes break down the pectins holding the cells together. Fans may call the flesh creamy and flavoursome, but I'm not convinced. For me, an apple needs to have some crunch.

Composting lies

Do you have a compost heap? Or maybe a compost bin or a compost pile or a fancy compost tumbler? I have tried a variety of different composting gadgets that have all resulted in either a heap of dried-out twigs or a horrible slimy mess. I have ended up, after many years of practice, with a couple of large black plastic compost bins. From these I can just about produce something that passes as compost, as long as you don't look at it too closely. Despite my most diligent efforts, it doesn't resemble the stuff you buy from the garden centre. Yet if you look online, or watch garden programmes, or read gardening magazines, you will be repeatedly and emphatically told that composting is easy and that we should all do it. The first of these statements is clearly and demonstrably untrue, whereas the second is definitely true.

Let's make a conservative estimate that each week the average household in the United Kingdom produces 2 kg (4½ lb) of compostable waste, such as banana skins, vegetable peelings, apple cores and so on. Multiply that up and you are looking at in excess of 5 million metric tonnes (nearly 5½ million US tons) of waste every year. Of course, this doesn't include wasted food each year (8 million tonnes, or 9 million US tons) and the annual debris from looking after a garden (4 million tonnes or 4½ million US tons). Put together, that's 17 million tonnes (19 million US tons) annually that many of us could be recycling ourselves, if only we had a reliable compost system.

A compost heap works because of certain bacteria and, to a lesser degree, fungi that break down the vegetable matter. To make good compost you need to create an environment that keeps these bacteria happy. There are a number of things you have to get right, and if you do, you'll produce the dark brown, crumbly and pleasant-smelling mixture that gardeners seek.

The first thing to get right is that your compost heap should be made up of 1 part nitrogen to 30 parts carbon. Which is gardening reference book jargon for 1 part leafy green stuff to 30 parts stalks and other tougher plant material. If you put too much nitrogen, or leafy stuff in the heap it will become a slimy, unpleasant mound. Too much carbon and you end up with a dry pile of sticks. Unfortunately, without a high tech lab in your shed, it's impossible to measure this ratio. Instead, you need to work from experience and guesswork, based on an understanding of plant science. Take for example grass cuttings from a lawnmower, made up entirely of leafy green stuff. Taken alone, the lawnmower's contents contain too much nitrogen for good composting. On the other hand, autumn leaves are very poor in nitrogen, as they are really just the dried-up husks of fleshy green leaves. Whenever you add something to the compost, you need to consider the balance of nitrogen to carbon. If you are about to chuck a load of grass or vegetable peelings in the compost, you probably need to add a bit more carbon in the form of torn-up paper or sawdust. Conversely, if you have just pruned back a load of twiggy material, this will be too carbon rich, so you need to add some leafy stuff.

Assuming you get your carbon to nitrogen correct, you then need to make sure that the pile is neither too wet nor too dry. In the United Kingdom, the problem is rarely too dry, but you may need to cover the compost to stop it becoming waterlogged in winter. Alongside the water issue is the need to keep the compost aerated so that the bacteria can get to oxygen. If your pile is too wet or too compacted there will be no oxygen and no bacteria. The traditional solution is to laboriously turn your compost, but recent research shows that you can just use scrunched-up cardboard, toilet rolls tubes or anything compostable that creates air spaces.

Another element that is often overlooked is that the bacteria that thrive in a compost heap prefer a slightly alkaline environment, and are definitely not happy if the heap is acidic. Once again, short of actually breaking out the science test kit, it's tricky to know if this is the case. However, if your compost is failing, despite your best efforts, it may be worth chucking on a bit of gardeners' lime, which is really just powdered limestone, or you could use wood ashes. Both of these will neutralize the acid, making the heap slightly alkaline, and creating the best environment for your bacteria to work their magic.

Finally, you need heat. As the bacteria get to work they start to kick out heat and the compost can, in theory, reach temperatures of up to 70 °C (158 °F). This high temperature will help bacterial growth and will also kill other organisms in the heap, including weed seeds. Unfortunately, achieving such high temperatures is really difficult in a domestic compost pile. To attain them you usually need huge mountains of compost that insulate the insides of the pile. Assuming you don't have a mountain of compost, the advice is to place your pile in the sunniest spot in the garden. Which is not particularly helpful, as the usual spot for a compost heap is behind the garden shed in a gloomy corner, where nothing grows and nobody wants to go. Failing the use of a sunny spot, you could just insulate your compost with old carpets, or something such as straw bales.

On the surface then, to make perfect compost you need to get your carbon to nitrogen ratio correct, make sure it's all well aerated, not too wet or dry or too acidic, and build a really huge compost heap. All of which makes it seem like a rather daunting prospect, but this fails to take account of the huge benefits that composting brings. Not only does it mean you have a supply of your own compost, freeing you from lugging bags of the stuff home from the garden centre, but it has a wide-ranging ecological impact. By composting, you will be significantly reducing your contribution to land fill, not only in terms of volume of material but also in methane emissions from piles of rubbish. In your garden, you will be improving the fertility of your own soil by recycling nutrients back into it, and giving the biodiversity of your patch of planet Earth a massive boost.

From my own experience and struggles with composting I know that while it's difficult to get perfect, it turns out that the worms and plants don't care if your compost is a bit rough and ready – it will still do the job.

The lure of the light

That moths and various other nocturnal flying insects fly towards a light is well known to us all. You leave a window open in summer with a light on inside and before you know it there are half a dozen winged beasties circling and bouncing off the light. But if you pause for a moment and consider what is going on, it becomes apparent that this is extremely bizarre behaviour. The insects that have gathered around the source of illumination are all night-flying creatures that make their living zooming about in the dark. During the light of day they creep into a corner or under a leaf, and hide from the keen-eyed predators that would make lunch of them. So why on Earth do they head straight for the first light bulb they see? The light doesn't indicate the presence of food, and it puts them at risk of being gobbled up by canny predators.

The most common explanation given for this behaviour is to do with moon-aided navigation. The idea here is that for a moth to find its way around at night, it takes a bearing on the Moon and keeps the Moon at a constant angle to the direction of its flight. Since the Moon is so very far away from the moth's flight, the angle does not change and it is an effective landmark. The moth quite reasonably confuses a light bulb for the Moon. As it flies past the illuminated bulb, it takes a bearing on the artificial light to keep it in a straight line. Unfortunately, as the moth gets nearer to this light, the angle between it and the moth changes. So, our moth corrects its course, steering back towards the bulb. This continues to happen and the moth circles the light, getting closer and closer until – whack! – it hits it. The lunar navigation of the moth evolved with an incredibly distant light source, and not our pesky close-up ones.

Regrettably, there are a couple of problems with this explanation. Firstly, most moths do not habitually go for long flights in straight lines. Many moths live in sheltered wooded environments where they just flit small distances and never really see the Moon. Yet these moths are still attracted to light. Also, when moth scientists looked closely at the behaviour of moths near artificial lights, many moths ploughed straight into the lights without the circling behaviour that the moon navigation theory predicted.

Another slightly kooky idea came from a 1970s entomologist at the US Department of Agriculture. It had been discovered that some of the super sexy pheromones released by female moths, that did a wonderful job of attracting male moths, were ever so slightly luminescent. The pheromones emitted light in the infrared spectrum that matches some of the light coming from a candle. The argument put forward was that male moths are fatally attracted to a flame because they think they are in line for an amorous encounter. But again, observations in the field make a mockery of this explanation. The best way to attract moths in the wild is to set up ultraviolet lights, not infrared ones.

There is one explanation that relies on observation. If you disturb a bunch of moths at night, on a shrub for example, they fly up into the air to escape. They do not fly down into the shadow of the vegetation. It looks as though the escape reaction for moths is to head up, possibly towards brighter regions, rather than down into darker shadows. So, if moths at night are alarmed, maybe they fly towards the light to escape. Although if true it implies that all moths you see attracted to lights have been disturbed somehow from their regular nocturnal business.

Naturally, given the complexity of biology, the actual answer, which currently eludes us, will probably be a mixture of a number of reasons. What we do have a good theory for is why the moths hang around once they get into the light. Moth eyes are adapted for seeing in the dark, and once they find themselves in bright light their poor little eyes are overwhelmed by the light and struggle to cope. They are effectively blinded by the light, which may explain their reticence to leave, as they can't see where they are headed. Of course, another explanation could be that they are just taking a breather after bashing their head against a light that they can't help zooming around.

The shade of a tree

If you find yourself outdoors on a hot, sunny day, we all know that retreating into the shade can provide relief from the heat. However, the observant among you will have noticed that not all shade is equal. If you shelter from the Sun in the shade of a building or under an awning of some sort, the temperature will be lower, but not as low as if you retreat beneath the canopy of a tree. Measurements made in North American cities of shade temperatures under trees were up to 3 °C (6 °F) cooler than temperatures in the shade of buildings within the city. Which seems counter-intuitive, as shade from trees is usually dappled rather than complete. It seems from simple measurements that trees actively cool their immediate environment.

When I tell you that leaves on trees are green, at least for the majority of species and varieties, it is no surprise to anyone. However, the implication of this is possibly more surprising. The foliage on plants reflects most of the green portion of the light striking it, which is why we see it as green. Furthermore, if you go beneath the canopy of a tree and look up, what light that does penetrate is also green – only green light is transmitted through the leaf. What happens to the red, orange, yellow and blue light? The plant's leaves absorb these colours of light, but they are not converted into heat energy, instead the light is harvested by the plant to power the process of photosynthesis. This is the biochemical reaction that makes sugar by taking carbon dioxide from the air, water from the ground and energy

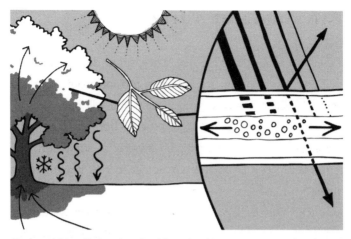

Red and blue light absorbed by a leaf is not converted to heat, but used for photosynthesis

from sunlight. This energy from the Sun is turned into chemical potential energy and stored for the plant to use later.

Now compare this to sitting under a green canvas canopy. In the same way as leaves, the green light is reflected, little gets through, and all the other colours of light are absorbed by the canopy. This time though, the light energy is turned into heat energy, half of which is radiated back up into the sky and the other half radiated down onto you sitting beneath the canopy. Even though the canopy is the same colour, since it is not actively storing the energy, you get hotter.

There is another, possibly even greater cooling effect of green plants. The process of photosynthesis, that uses the sunlight I just mentioned, needs water as a basic ingredient. To get this, the plant has a system of several mechanisms that draws water up from its roots to the tips of every leaf. Part of this process is known as transpiration, in which the plant allows water to evaporate from its leaves, and in so doing helps pull proportionally more water up from its roots. Some of this water is used for photosynthesis and some for more transpiration. One of the consequences of this is that the process of evaporation of water from the leaf draws in heat energy and the leaf cools. This is called evaporative cooling, and is also the process that allows refrigerators to work (see page 45). When this happens across the entire canopy of a big tree, the heavier, cool air around each leaf then falls below the canopy, making the space beneath the tree cooler.

Combined, the consequence of photosynthetic light use and evaporative transpiration can create a pleasantly cool place to sit. This cooling is an active process, and more than just an absence of direct sunlight. However, there is a bigger picture here, as a sprawling city centre can create an effect known as a heat island. Large masses of dark buildings and hard, reflective surfaces within a city centre can easily raise the temperature 6 °C (11 °F) hotter than the surrounding suburbs. City planners have found that one of the best ways to mitigate this effect is to plant trees and lawns; ideally not just at ground level, but also on top of buildings. The result can bring the overall temperature back down a bit, and also create even cooler localized areas where residents can relax, sit underneath a tree and enjoy their lunch.

Can a spider catch itself?

There is something deeply unnerving about spiders, and yet at the same time I find myself fascinated by them. When I see a big, bloated orb weaver in the middle of its web, I can't help but have a good, close look. Then when it scuttles off to enshroud an unfortunate victim, I find myself drawn to watch this gruesome scene play out. The victim of the spider is held, stuck fast in the web, while the spider dashes across the very same sticky threads. The intriguing science behind the spider's ability to do this is only now becoming apparent.

Spider silk is an incredible material with great future potential, once we unravel its secrets. It is stronger than steel and tougher than the bulletproof synthetic material Kevlar, yet a spider can produce metres (or yards) of the stuff on demand. What is more, the average spider produces seven different types of silk, each with its own unique properties and uses. Fortunately for us, we only really need to look at a couple of these to unstick the conundrum of how a spider manages to not catch itself in its own web.

The spokes of the web made by the garden orb weaver spider, *Araneus diadematus*, are made with what is known as major ampullate silk. This is the strongest silk that spiders produce, and the one about which we know the most. Significantly, it has no glue on it at all; the task of being sticky falls to the flagelliform silk. This type of silk forms the spiral around the spokes and is coated in regularly spaced droplets of sticky glue. Except of course for the group of spiders that don't use glue, but instead rely on a different type of silk and electrostatic forces, in the same way that geckos' feet stick to smooth surfaces. As you delve into the biology of spiders, it quickly becomes apparent that their silk comes in a myriad of forms with an incredible range of functions. Diversions aside, if we return to our garden orb weaver, the primary way that the spider does not get stuck in its own web is by not treading on the glue droplets. When a spider walks across its own web it's careful to only walk on the major ampullate silk that is not only stronger, but is also free of glue. Clearly, when an insect crashes into the unseen web, it has no such luxury of choice and encounters the gluey, as well as the glue-free, strands.

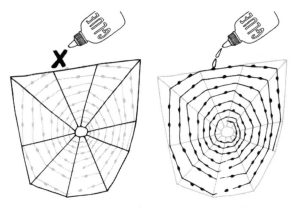

The spokes of a spider's web are not sticky but the spirals are

The spider has more tricks up its sleeve though. At the end of its main walking legs are a specialized set of tiny claws. Two of these are for gripping onto twigs and leaves, while the third is specially designed to hook over threads of silk. Nestled just beneath this third hook are a set of springy bristles. When the spider grabs onto a silken thread with the special third claw, the bristles are pushed back and held under tension by the thread. On release of the thread, sticky or not, the bristles push back and flick the thread away with enough force to overcome any glue.

Finally, the spider has another adaptation that helps specifically when it is building its web. The spider is forced to touch the sticky strands when it lays them out in its web. The tips of the legs it uses for this task, usually the back pair, are covered in short bristles with very fine points. There is also evidence of some sort of chemical coating or surface layer and, when taken together, the specially coated bristles repel the glue on the sticky threads.

When you are next watching the gruesome ballet of a spider running across its web to the struggling prey, consider that the simple answer to how it manages to not get itself caught is that it takes care to tiptoe around the glue. Which is probably much harder than it sounds when you have eight legs and can't see your own feet. However, if that fails and a spider absolutely must touch its sticky-trapped silk strands, it is armed with a battery of special adaptations to keep it free from its own web.

The impossible lawn

Do you have a lawn? Is it full of weeds? Have you spent time and money trying to kill off moss, creeping buttercups and dandelions? If the answer to the first is yes, then the answer to at least one of the other questions is almost certainly also yes.

The garden and parkland lawn did not really exist until the seventeenth century, when it became trendy among the British aristocracy. Out went complicated gravel paths and in came close-cropped grass lawns. The well-to-do of the era could pop out to enjoy the fresh air while walking across their small patches of lawn. Then, in the eighteenth century, along came the first garden design, or 'landscape architect', superstars, such as Capability Brown, who converted the fashion for fiddly, formal gardens into huge sweeps of pastoral fantasy. For terribly wealthy clients, Brown would create his trademark serpentine lakes surrounded by great swathes of lawn with deer and carefully placed sheep. The lawn was only to be found at the houses of the super-rich, simply because it took huge manpower to keep it cut using scythes and hand shears. Then in 1830, Edwin Beard Budding of Stroud, England, invented the lawnmower. It did not take long for the rising middle classes to grasp this new invention firmly by the handles and start the craze for domestic lawns.

The thing about maintaining a lawn is that the very action of cutting encourages grasses to proliferate. Trees, for example, do very poorly if you repeatedly and regularly cut them down just above the ground. In fact, many plants cannot cope with being regularly felled. The bit of a plant that does the growing is called a meristem, and in plants such as tulips, begonias and carnations, the meristems are at the tips of the growing shoots. If you cut these plants off at the ground, you chop off the meristem, and they have to start the process of producing a shoot all over again. This regrowth can only happen at considerable energy cost to the plant. If you keep doing this, eventually the plant will give up the ghost and die. On the other hand, grasses keep their meristems tucked away at the base of the plant. If you chop off the leaves of a grass plant it merrily continues to grow from the bottom, unconcerned by the decapitation it receives. Grasses have evolved this system in response to being repeatedly munched on by herbivores. By regularly mowing a lawn, the only plants that can survive are grasses.

There are some significant exceptions to this rule. Any plant that grows very close to the ground, keeps its leaves flat and never pokes itself very high will also thrive. Buttercups, daisies, dandelions, clover and moss are all perfectly happy living under the regime of close shaving that lawns undergo. This partly explains why your lawn is riddled with non-grass weeds, as there are many plants that will thrive in a lawn. But it is worse than that: you are fighting a biological battle that you will never win.

There is a relationship between the number of plant species you find in an area and the amount of biological material, or biomass, in that area. Ecologists have travelled all over the world and measured the number of species found in 1 m sq (about 10 sq ft) patches and also the biomass in the same patch. While it may be a leap from natural habitats to your artificially managed lawn, it turns out that the amount of biomass in a garden lawn should correspond to the very peak of the number of different species. Your lawn is biologically perfect for incredible biodiversity. Short-cropped turf on limestone, for example, is some of the most botanically diverse land in the world. It averages forty different species per square metre (about 4 per sq ft), and this richly varied plant habitat is what your lawn would naturally become.

The dream of a lawn made solely of closely cropped, green grass is biologically against all the odds. There are only two solutions to this dilemma. You could use herbicides and endless raking and weeding to keep your grass weed free, or you could embrace the biological imperative of diversity. After all, even if it's not grass, it's still green, and there is the added benefit that if you leave it long enough it may also turn yellow, white, pink and purple.

The colour of autumn

One of the greatest joys of autumn is enjoying a woodland walk surrounded by the yellows, golds and reds of the leaves on the trees and kicking your way through the leaves on the ground. Curiously, for something so familiar to us all, it is not at all clear why this happens.

It is worth pointing out that leaves don't drop to the ground because they are dying – rather, the tree initiates an active process of clever recycling called senescence. A tree, like an oak for example, would struggle to survive through a harsh winter if it retained its canopy of leaves. It would risk damage from strong winter winds and would lose more water from its leaves than it could draw up from the frozen ground. If it didn't blow over, it would die of thirst. As winter approaches, the length of the day shortens, the temperature drops, and plants, including trees, can detect this change. It signals to them that it is time to shed their leaves. First, however, trees carefully suck all of the useful nutrients out of the leaves and then, with surgical precision, block up that pathway into the leaves. That blocked pathway at the base of the leaf stem creates a weakness and, in the wind, the leaves snap off and fall to the ground.

One of the key substances that trees scavenge back from their leaves is the chlorophyll that gives the leaves their green colour. This contains valuable magnesium, so is worth the effort of salvaging for the plant. Chlorophyll is what enables all plants to capture the energy of sunlight, but it's not the only pigment to be found in tree leaves. Leaves are also often full of yellow and orange pigments called carotenoids – and yes, such pigments are what make a carrot orange. These carotenoids help the chlorophyll capture sunlight, but you don't normally see them under the green colour. What's more, since they don't contain any useful minerals, plants don't bother salvaging the carotenoids from their leaves. As the leaves are stripped of chlorophyll they gradually turn from green to yellow or orange.

This explains why you see yellow autumn leaves on birch trees, but not why leaves on a maple tree turn red. That red colour comes from pigments called anthocyanins, which are not found in green leaves. Trees with red autumn colour manufacture anthocyanins at the same time as they pull the chlorophyll out of their leaves. This is an odd thing to do, as the red colour costs the plant energy to make, and is lost in the shed leaves. The tree must get some benefit from the red colour, as sadly it does not just do it for our enjoyment.

Three theories have been put forward for the costly production of red pigment in autumn leaves. The first relies on the antioxidant properties of anthocyanins. Ironically for plants, one of the toughest things they have to cope with is sunlight. In green leaves the chlorophyll does a good job of soaking up the Sun's energy, but with the chlorophyll removed the same energy can wreak havoc. It knocks electrons off molecules, turning them into free radicals; these super-reactive molecules cause all sorts of damage inside plant cells unless mopped up by antioxidants. The red colour we see may function like a sunscreen, allowing the leaves to remain useful right to the end of their working life.

This explanation cannot be the whole story, as the autumn display of fiery reds, so familiar in North America, is replaced in Europe with a range of mostly oranges and yellows. This is where the humble, sap-sucking aphid may have played a part. Studies of autumnal orchards have shown that trees that turn red seem to suffer from fewer aphids than yellow-coloured trees. Possibly, red leaves are unappetizing to aphids, because they are chock-full of nasty-tasting anthocyanins. This theory goes on to suggest that during the last Ice Age, insect populations were trapped and killed by the ice sheets that developed from the north and Alpine south. European trees then evolved without so many wintering bugs from which they needed to manufacture anthocyanin protection, and lost their autumnal red colours. In America, no such icy pincer attack occurred, so the red colour remains.

The last of the currently held possible reasons for the red autumn colour is positively sinister. In one study, botanists found that all the anthocyanins produced and then discarded by a tree had a toxic effect on nearby saplings. It would appear that the gorgeous red colour of maples may be the tree's way of poisoning its competitors.

At the moment we don't really know all the reasons for autumn colour, and it probably arises for a combination of reasons, none of which prevents us from simply enjoying the spectacle.

Lunacy and the distant sound of thunder

My all-time favourite scientific hero was the eighteenth-century physician, poet, natural philosopher and polymath

Erasmus Darwin, who also happened to be Charles Darwin's grandfather. Erasmus Darwin lived in Lichfield, north of Birmingham in England, and his home there became a centre for free thinking and scientific enquiry. Among the many activities in which he became involved was the Lunar Society, so called because its meetings were held on the night of the full moon, in order that there would be light by which the carriages of the members could return home. If you peruse the membership of the society, it reads like *Who's Who* of eighteenth-century science. It included such luminaries as James Watt, of steam engine fame, Joseph Priestly, discoverer of oxygen and inventor of fizzy drinks, and even, occasionally, Benjamin Franklin. These great men of science would meet for dinner to discuss the latest topics of natural philosophy, drink port and, crucially, perform experiments.

It was one such experiment that they attempted to carry out that was aimed at a subject I often ponder when I hear it: why does thunder rumble? Following a flash of lightning, you normally have a couple of seconds in which to prepare yourself for the rumble of thunder. In that time, it's worth focusing your attention on the precise nature of the sound. What you usually hear is a loud crack followed by a rumbling noise of fluctuating volume. The rumble can go on for tens of seconds and can be as loud as the initial crack.

Back in the eighteenth century, Erasmus Darwin and his fellow 'lunarticks' decided to investigate what was going on with the rumble of thunder. They knew that the sound was instigated by the flash of lightning, although not exactly

Inattentive 'lunarticks'

why or what made the rumble. So, they concocted a scheme to create an artificial bang, high up in the air. One of the members, Matthew Boulton, an industrialist from Birmingham, constructed a huge paper balloon 1.5 m (5 ft) in diameter. This balloon was duly filled with a lighter than air, and highly explosive, mixture of hydrogen and oxygen. To this floating bomb they attached a fuse, then lit the fuse and released the balloon to float into the evening sky.

Unfortunately, the fuse they attached burnt much more slowly than expected and as they waited with, no doubt, a glass of port in hand, they got distracted and started chatting. When the balloon did go off with a colossal bang, they were so surprised that they forgot to listen for any rumble. Fortunately, James Watt had not attended this meeting, and heard the bang from his house nearby. He reported that after the initial bang, the noise had rumbled for an additional second or so. The conclusion drawn by the Lunar Society was that thunder rumbled because the sound of the lightning was being reflected from nearby hills.

Thunder does indeed rumble because of reflection from nearby building and topographical features, but this is only a minor element. What the Lunarticks failed to do was to simulate a bolt of lightning.

When lightning strikes, the flow of electricity through the air creates temperatures hotter than the surface of the Sun, in excess of 20,000 °C (36,000 °F). This super-heated air expands at an incredible speed and hits the surrounding cooler air with a powerful shockwave. Since the process happens at supersonic speeds, the shockwave creates a sonic boom. The crucial thing that Darwin and his friends missed is that this happens along the entire length of the lightning bolt.

Imagine a hypothetical bolt of lightning that streaks vertically down from the sky and hits the ground about 2 km (1¼ miles) away from where you stand listening. Now consider that the average lightning bolt is 10 km (6 miles) long, so while the bottom of the bolt is only 2 km away, the top is over 10 km away. If you take into account the speed of sound, it will take the sonic boom from the base of the strike only 2 seconds to reach your ears. However, the noise from the very top of the bolt, assuming you can hear it at the distance, will take up to 30 seconds to get to you. The intervening period of time will be filled with the rumbling boom coming from all the bits of the lightning bolt between the bottom and the top.

Lightning is rarely, if ever, so neat and tidy as our hypothesized bolt. It will switch directions, to create the archetypal zigzag pattern, so that the middle of the bolt could be further away than the top. You also often get lightning that jumps between clouds, travelling horizontally and never hitting the ground. If this sort of bolt is going directly away from you, the start of it will be further away than the end of the strike. All of this, coupled with sound reflections, will complicate what you hear, and create the rolling rumble of thunder.

So, with this understanding, we can now redesign the experiments of Erasmus Darwin and his fellow lunarticks. Clearly what is needed is a chain of balloons, filled with explosive gas that stretches kilometres into the sky. This should give a suitable rumble of thunder, although I don't fancy filling in the risk assessment for this one.

My rainbow, not yours

The first thing I do when I see a rainbow is look for a second. Now this may seem a trifle greedy of me – after all, surely one such marvel of nature is enough? However, a single rainbow can be an opportunity to observe a whole panoply of gorgeous science, including secondary rainbows, an Alexander's band, and even supernumerary rainbows, if you're really lucky.

First, I need to go over the essentials of rainbology – which is a word I just made up, in case you were wondering. For a chance to see a rainbow, you need two things; sunshine and rain showers. You don't need to stand in sunshine yourself,

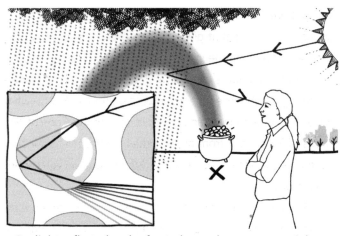

Sunlight reflected and refracted to make your own rainbow

but the sunlight must be shining onto an area of rainfall that you can see. Since this requires that the Sun shines down at a fairly shallow angle, you are most likely to see rainbows in the morning or evening, or in the winter and spring. The area of rain also needs to be directly in front of you when you have your back to the Sun.

The need for these convoluted arrangements of yourself, the Sun and the rain, is that a rainbow is really a complicated experiment in optics. The light going into your correctly positioned patch of raindrops bounces once off the back of the raindrops and comes back out at the front. The reason why the colours appear is down to the process of refraction. When light passes from air into water it slows down a little bit and this makes it change direction ever so slightly. The amount of this bend depends on the wavelength or colour of the light, with red light bending the least and violet light the most. As the light enters the raindrops it bends and splits into different colours and it's this that gives us the rainbow, with red on the outside and blue on the inside of the bow.

This has a couple of implications that still give me pause for thought whenever I encounter a rainbow. Firstly, what you see as a rainbow is actually built up from millions of tiny points of light, each from an individual raindrop. The rainbow is effectively pixelated, but at such a high resolution you cannot make it out.

Secondly, the position of a rainbow in the sky depends on where the viewer is standing. The reflections off the back of each raindrop only reach your eye if the angles are all exactly spot on. If you take a step to one side you see light reflected from different raindrops in different positions. A consequence of this is that we each see our own, unique rainbow. It is quite possible for you to see a full rainbow while somebody else standing near by only sees a partial one. To take this to its logical extreme, each of your eyeballs sees a different rainbow. Pause for a moment to consider this, and I challenge you next time you see a rainbow to try shutting each of your eyes in turn to see if you can spot the difference.

Astonishingly, rainbows were pretty much fully explained at the end of the thirteenth century. The research was independently completed at about the same time by both an Iranian scholar, Kamal al-Din al-Faris, and a German Dominican friar known as Theodoric of Freiburg. Both men used spherical glass flasks to show the path of light inside a raindrop, presumably making their own tiny rainbows.

The other name that often crops up with rainbows is that of Isaac Newton. By the middle of the 1600s, we still didn't understand where the colours of the rainbow came from. If you take a glass prism or a water-filled sphere you can make a rainbow. White light goes in, colours come out. To explain this there were two competing ideas: either the prism or sphere coloured the light somehow, or white light was made up of all the colours. On the face of it, neither seems a particularly likely explanation. Then in 1666, while Newton was at home in Woolsthorpe, Lincolnshire in the UK, he performed his *experimentum crucis* or crucial experiment. He took a ray of sunlight and split it into a rainbow with a prism. He then took a lens to gather the rainbow light back together and focused it onto another prism; this time, white light came out of the other end. Thus, white light is made of a mix of coloured light. We know he did this experiment in Woolsthorpe because of notes he made in one of his journals. In these notes he lists the distance from the hole in his shutter through which the ray of sunlight shone to the far wall. I've measured his room in Woolsthorpe and it's spot on. At the time, I was told by the National Trust curators that I was the first to do this, but I don't believe it. Still, it's a good story for the kids.

As an aside, Newton named the peculiar colours of the rainbow that we have. I mean, can you tell the difference between indigo and violet? Initially he had just five: red, yellow, green, blue and violet. Only later did he add orange and indigo to the list. Significantly, this gave him seven colours and that suited his penchant for alchemical numerology: seven colours for the seven musical notes and the seven planets that he knew about.

For those of you who like myself can't get enough of a good thing, where you find one rainbow you often find two. Next time you see a rainbow, take a look a little bit further out from the main, or primary rainbow. If you're lucky you should see a second rainbow. Normally a secondary rainbow is more diffuse and very, very faint. So faint that people usually fail to notice it, but it's often there. It's caused by exactly the same process as the main rainbow, except that the light bounces twice inside each raindrop. This second bounce inside the drop flips the colour sequence of a secondary rainbow, red on the inside of the bow and violet on the outside.

While it may be greedy to want two rainbows, I love the added excitement of discovering this secondary rainbow. What's more, it need not end here. There are a couple of other obscure bits of rainbology that are worth looking out for. The first, called the Alexander's band, named after an Ancient Greek philosopher, is the noticeable darkening of the sky between the primary and secondary rainbow. Once you have seen this, keep an eye out for what are known as supernumerary rainbows. These are little stripes of colour, usually green and blue, on the inside of the main primary rainbow. Both of these effects are caused by complicated optical reflections and interferences that you only see when the raindrops are of a very uniform size and distribution. Which is why they are particularly rare, as finding the conditions just right to see either the Alexander's band or supernumerary rainbows is an unlikely event.

At the heart of all rainbology is the deeply satisfying idea that the rainbow you see is yours and yours alone. Since it all comes down to being at precisely the right angle to the rain and Sun, all of the secondary effects are also unique to you. What's more, if you are always at the centre of your own rainbow, you can never get to the end of this rainbow. While this puts paid to any Leprechaun-based ambitions, you still get to marvel at your own version of nature's most impressive lighting display.

Acknowledgements

The book in your hands is only here due to the efforts of a number of very important people. Firstly my agent, Sara Cameron, who manages to somehow know everyone and seemingly mentions my name to them all. For this I am eternally grateful as it has given me an opportunity to write this book.

I must also thank the lovely people at Michael O'Mara Books, in particular Hugh Barker and my editor Gabby Nemeth, who have helped ease the process of the writing and, with humour and patience, dealt with my fumblings and ramblings and odd questions. Although I am still not sure at what point you stop converting to imperial units. Are micrometres small enough, or do we go all the way down to nanometres?

Finally, and most importantly I have my wife, Juliet, to thank for keeping me, as always, on the straight and narrow. Without her editing, deep understanding of science and companionship this would be a very different piece of writing. Also, and this pains me to admit, I would not have hit any of my deadlines without her help.

Index